Social Media, Sociality, and Survey Research

Social Media, Sociality, and Survey Research

Edited by

Craig A. Hill
Elizabeth Dean
Joe Murphy

RTI International
Research Triangle Park
North Carolina

For general information on our other products and services or for technical support, please contact our Customer Care Department within the United States at (800) 762-2974, outside the United States at (317) 572-3993 or fax (317) 572-4002.

Wiley also publishes its books in a variety of electronic formats. Some content that appears in print may not be available in electronic formats. For more information about Wiley products, visit our web site at www.wiley.com.

Library of Congress Cataloging-in-Publication Data

Social media, sociality, and survey research / [edited by] Craig A. Hill (RTI International, Research Triangle Park, NC), Elizabeth Dean (RTI International, Research Triangle Park, NC), Joe Murphy (RTI International, Research Triangle Park, NC).
    pages cm
  Includes bibliographical references and index.
  ISBN 978-1-118-37973-8 (cloth)
  1. Online social networks–Research.   2. Social media–Research.   3. Social surveys–Methodology. I. Hill, Craig A., 1956- II. Dean, Elizabeth, 1974- III. Murphy, Joe, 1974-
  HM742.S62845 2013
  302.23′1–dc23

                                         2013018926

Printed in the United States of America

10 9 8 7 6 5 4 3 2 1

Contents

List of Figures

List of Tables

Contributors

Sarah Cook, RTI International, 3040 Cornwallis Road, Research Triangle Park, NC 27709

Elizabeth Dean, RTI International, 3040 Cornwallis Road, Research Triangle Park, NC 27709

Jill Dever, RTI International, 701 13th Street, N.W., Suite 750, Washington, DC 20005-3967

Carol Haney, 75 West Street, Dummerston, VT 05301

Heather Hansen, RTI International, 701 13th Street, N.W., Suite 750, Washington, DC 20005-3967

Saira Haque, RTI International, 230 W Monroe Street, Suite 2100, Chicago, IL 60606

Brian Head, RTI International, 3040 Cornwallis Road, Research Triangle Park, NC 27709

Craig Hill, RTI International, 3040 Cornwallis Road, Research Triangle Park, NC 27709

Michael Keating, RTI International, 3040 Cornwallis Road, Research Triangle Park, NC 27709

Annice Kim, RTI International, 3040 Cornwallis Road, Research Triangle Park, NC 27709

Joe Murphy, RTI International, 230 W Monroe Street, Suite 2100, Chicago, IL 60606

Rebecca Powell, Survey Research and Methodology, 201 North 13th Street, Lincoln, NE 68588-0241

Jon Puleston, GMI, 1 Bedford Avenue, London WC1B 3AU, United Kingdom

Bryan Rhodes, RTI International, 3040 Cornwallis Road, Research Triangle Park, NC 27709

Ashley Richards, RTI International, 3040 Cornwallis Road, Research Triangle Park, NC 27709

David Roe, RTI International, 3040 Cornwallis Road, Research Triangle Park, NC 27709

Adam Sage, RTI International, 3040 Cornwallis Road, Research Triangle Park, NC 27709

Jodi Swicegood, RTI International, 3040 Cornwallis Road, Research Triangle Park, NC 27709

Yuying Zhang, RTI International, 3040 Cornwallis Road, Research Triangle Park, NC 27709

Preface

We started with the premise that survey researchers should be thinking always about the future. That has never been more true than now. Beset by problems not of our own making, the survey research discipline faces unprecedented challenges as a result of declining data quality stemming from, for example, falling response rates, inadequate sampling frames, and approaches and tools that have not adapted to the rapid pace of technological change, especially the changes in the way human beings (our respondents) communicate with each other. Conducting a survey is, at its core, a social interaction between a researcher (represented by an interviewer, or, increasingly, a computer screen) and a (potential) respondent. Yet, the current pace of technological change—and the way people communicate with each other—threatens the upheaval of survey research as we know it because people expect modern communication to take place differently than it did when we developed the current set of best practices for survey research.

Thus, survey researchers should—and must—search for ways to improve the manner in which research is conducted. Survey researchers should—and must—constantly scan the landscape of technological and social change to look for new methods and tools to employ. In this spirit, we have been somewhat jealously watching the explosion of "social media." Social media is, no doubt, having a profound impact on communication styles and expectations, and thus, more than likely, will have an equally large impact on the way we conduct social science.

We have organized this book around the idea of the "sociality hierarchy"—that is, there are three "levels" of sociality inherent in the current (and future) state of person-to-person interactions using computing devices: (1) Broadcast, (2) Conversational, and (3) Community. Survey researchers should recognize these levels when attempting to apply new social media tools to survey research. We show examples of how this can be done and, perhaps more importantly, how survey researchers should think about applying these in the future as a complement to "traditional" survey research.

Chapter 1 discusses the advent of social media in its many and varied forms and defines it from the perspective of a survey researcher. We also show why survey researchers should be interested in, and vigilant about, social media—and the data it produces. We introduce the concept of the sociality hierarchy for social media and show examples of each level or category.

Chapters 2 and 3 examine broadcast-level social media—the first level in the sociality hierarchy. In Chapter 2, Haney provides a handbook for sentiment analysis, paying close attention to the pitfalls that await researchers who do not consider carefully their research question. In Chapter 3, Kim et al. perform a case study of Tweets on health-care reform to determine whether such analysis could ever replace opinion polls on the topic.

Chapters 4 to 7 present examples of use of the conversational properties of social media for survey research, which is the second level of the sociality hierarchy. In Chapter 4, Sage describes using a Facebook application to build a sample of respondents and collect survey data by conversing with them in the special world of Facebook. In Chapter 5, Dean et al. demonstrate how researchers can use the virtual world Second Life and Skype videoconferencing software to conduct cognitive interviews (a conversation between researcher and participant) with a geographically dispersed population. In Chapter 6, Richards et al. use Second Life as a survey laboratory to test comprehension and compliance with the randomized response technique, a method of increasing the privacy of sensitive questions in surveys. In Chapter 7, Roe et al. describe the processes involved and decisions made in building a mobile survey panel that will, again, enable direct one-to-one "conversations with a purpose" between researchers and respondents.

Chapters 8 to 11 examine the community level of the sociality hierarchy. In Chapter 8, Keating describes how crowdsourcing techniques can be used to supplement survey research. In Chapter 9, Richards et al. present a method for using Twitter to collect diary data from specific Twitter-friendly communities. In Chapter 10, Haque et al. use extant social networks in Second Life to recruit and interview subjects with chronic medical conditions. Finally, in Chapter 11, Puleston describes methods for gamifying surveys—making them more interactive, interesting, and fun for respondents and, in effect, building communities of eager survey participants.

In the last chapter (Chapter 12), we use the sociality hierarchy to think of ways to improve the survey research of the future.

We consider ourselves fortunate to be employed by RTI International—an institute that considers innovation to be one of its core values. To foster this innovation (in survey research and all branches of science), RTI makes available internal funding designed to advance the science in the many disciplines under its roof. The great majority of the chapters in this book are tests, ideas, and experiments funded through this program, and we are eternally grateful to RTI for that opportunity.

Further examples and insights from this research are routinely shared in our blog SurveyPost (http://blogs.rti.org/surveypost), where readers can review, comment, and discuss topics germane to the future of survey and social research. One advantage of SurveyPost, of course, is that it allows us to respond quickly to new developments in technology and communications and survey research. This book has taken a comparatively longer length of time and, as a result, may contain some references that are out-of-date before it is printed, solely because of the speed of change in the social media and communications world.

Our intended audience for this book is the survey research community. However, our own backgrounds are quite diverse, having come to survey research from several different vectors and disciplines, so we hope that the book has broad appeal and finds interest among sociologists, political scientists, and psychologists, as well as those from the communications field, human–computer interaction researchers, market researchers, and all interested in the conduct of social science, both now and in the future.

CRAIG A. HILL
ELIZABETH DEAN
JOE MURPHY

April 2013

Acknowledgments

We would like to thank and acknowledge the following people for their contributions to this book: Ian Conlon, for review and comment on draft chapters; Laura Small, for copyediting support; John Holloway, for graphics coordination; Rebecca Powell, for data management and data analysis assistance; and Jodi Swicegood, for literature review assistance. We also want to thank RTI International for its support.

Tayo Jolaoso (RTI) developed the sociality hierarchy graphic in Chapter 1. In Chapter 3, the Kaiser Family Foundation provided public opinion poll data. Sarah Cho, Liz Hamel, and Mollyann Brodie of the Kaiser Family Foundation provided valuable feedback on an early draft of the chapter. The authors of Chapter 5 would like to acknowledge contributions made by Fred Conrad, of the Joint Program in Survey Methodology, to the avatar similarity instrument content and cognitive interview study design. The authors of Chapter 6 would like to acknowledge the vital work of John Holloway, who designed and developed the coin toss mechanism that was used in Second Life.

Social Media, Sociality, and Survey Research

Joe Murphy, Craig A. Hill, and Elizabeth Dean,
RTI International

As survey researchers, we have long been concerned about the future of survey research. Beset by problems not of its own making, the survey research discipline faces unprecedented challenges because of declining data quality—stemming from, for example, falling response rates, inadequate sampling frames, and antiquated approaches and tools. Conducting a survey is, at its core, a social interaction between a researcher and a (potential) respondent—a "conversation with a purpose." The current pace of technological change—and the way people communicate with one another—presages the upheaval of survey research as we know it.

Thus, survey researchers should be—must be—searching for improvements to the way their research is conducted. Survey researchers should be—must be—constantly scanning the landscape of technological and social change, looking for new methods and tools to employ. In this spirit, we have been, somewhat jealously, watching the explosion of social media.

In this book, we introduce the concept of the *sociality hierarchy*; that is, three "levels" of sociality inherent in the current (and future) state of

Social Media, Sociality, and Survey Research, First Edition.
Edited by Craig A. Hill, Elizabeth Dean, and Joe Murphy.
© 2014 John Wiley & Sons, Inc. Published 2014 by John Wiley & Sons, Inc.

person-to-person interactions using computing devices: (1) broadcast, (2) conversational, and (3) community. Survey researchers should recognize these levels when attempting to apply new social media tools to survey research. This book presents examples of how this application can be done and, perhaps more importantly, how survey researchers should think about applying these tools in the future as a complement to traditional survey research.

In this first chapter, we discuss the advent of social media in its many and varied forms, and we define it from the perspective of a survey researcher. We also show why survey researchers should be interested in, and vigilant about, social media. We introduce the concept of the sociality hierarchy for social media and show examples of each level or category.

Throughout the rest of the book, we explain (and show), at a more practical level, how survey researchers can use the data generated by social media at each level of the sociality hierarchy. Finally, we suggest particular vectors on which survey researchers might find themselves with regard to the use of social media data and tools as we move inexorably into the future.

WHAT IS SOCIAL MEDIA?

Millions of people have joined networks like Facebook and Twitter and have incorporated them into their daily lives, and at least partially as a result, communications between individuals and groups have changed in a fundamental way. Every day, billions of transactions occur over electronic systems, and within this stream are data on individuals' behaviors, attitudes, and opinions. Such data are of keen interest to those conducting survey research because they provide precisely the types of information we seek when conducting a survey.

The term *social media* has become ubiquitous. But what is social media? The term *social* suggests two-way interactions between people, which may be classified as one-to-one, one-to-many, or many-to-many. *Media*, or tools that store and deliver information, typically include materials that deliver text, images, or sound, i.e., *mass media* like books and magazines, television, film, radio, and *personal media* like mail and telephone.

Media can be further delineated into *analog media*, which contain data in a continuous signal or physical format, and *digital media*, which store information in a binary system of ones and zeros. The important distinction for our purposes is that digital media are typically stored or transmitted through computers or digital devices and can be disseminated via the Internet (aka the web).

The term *social media* most commonly refers to web-based technologies for communication and sharing over the Internet. There is no single agreed-upon definition of social media, but Scott and Jacka (2011, page 5) contend that social media "is the set of web-based broadcast technologies that enable the democratization of content, giving people the ability to emerge from consumers of content to publishers." Social media involves the intensive use of electronic media for people in contact through online communities (Toral et al., 2009), but no agreed-upon definition exists for the concept of "online community" either (De Souza & Preece, 2004).

We propose a specific working definition of social media for the purposes of survey research: *Social media* is the collection of websites and web-based systems that allow for mass interaction, conversation, and sharing among members of a network. In this definition, social media has four defining characteristics: user-generated content, community, rapid distribution, and open, two-way dialogue (Health Research Institute, 2012).

Social media must be distinguished from other similar terms that may refer more to the technological or structural aspects of online systems. For instance, the *web*, built on the infrastructure of the Internet, contains myriad sites that are not part of the social media construct. These sites include, for instance, information resources without an interactive, discussion, or sharing component. Such resources typify *web 1.0*, which describes resources that allow users to view and consume information from the web but without necessarily sharing or interacting with the contents (Krishnamurthy & Cormode, 2008). *Web 2.0*, however, refers to sites with user-generated content such as videos, music, blog text, and photos (Anderson, 2007). Web 2.0 has allowed users to interact with the web and with other users and has permitted the consumer to become the creator of content (Asberg, 2009). Web 2.0 has been used as an umbrella term for web-enabled applications built around user-generated or user-manipulated content (Pew Internet &

American Life Project, 2011). Ravenscroft (2009) considers web 2.0 to be the "social and participative web" that includes tools emphasizing social networking (e.g., Facebook, Bebo, and LinkedIn), media sharing (e.g., MySpace, YouTube, and Flickr) and virtual worlds (e.g., Second Life).

The distinction between social media and web 2.0 can be rather nebulous, but Figure 1.1 shows how social media fit into the evolution of popular media and technology. The figure presents analog, digital, web 1.0, and web 2.0 media types progressing over time from left to right. Analog media are listed with arrows connecting them to digital and web media that have evolved from or supplanted them. Diamonds represent media that have traditionally been employed for survey research including the following:

- Mail, used for delivering paper-and-pencil surveys
- Telephone, used for conducting survey interviews
- Digital audio, such as WAV files, used for recording interviews for verification and data quality purposes (Biemer et al., 2000)
- E-mail and homepages (now more commonly referred to as webpages) for web surveys, notifications, and panel maintenance
- Market research online communities, which are private social networks and websites to conduct qualitative marketing research with a selected community of members.

Media represented by hexagons in Figure 1.1 hold potential for survey research innovation moving forward and are, thus, the focus of this book. In this book, we present our research on these new modes and methods such as online games, multiply massive online games, and virtual worlds (Chapters 5, 6, 10, and 11); social networking platforms like Facebook and Twitter (Chapters 2, 3, 4, and 9); smartphone-enabled media like mobile apps for survey data collection (Chapter 7); and web 2.0-enabled research resources such as crowdsourcing (Chapter 8).

The dotted line in Figure 1.1 indicates which media we consider *social media* for the purposes of this book. It encompasses web 2.0 media but also includes some typically classified under web 1.0 that allow for user-generated content, rapid distribution, or open, two-way dialogue.

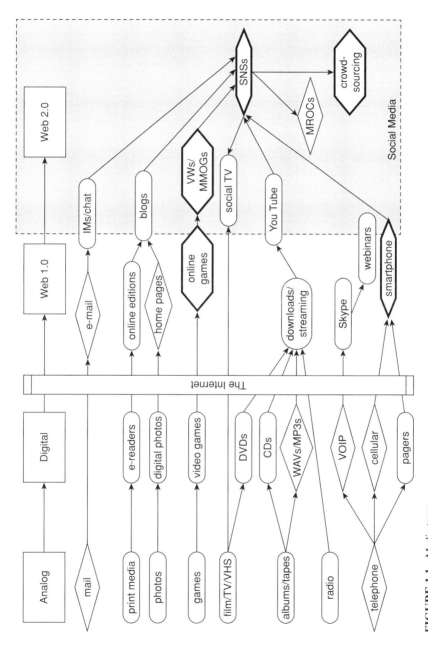

FIGURE 1.1 Media types.

5

SOCIAL MEDIA ORIGINS

Consonant with our definition of social media as web-based, most social media originated—were born—in the 1990s. The explosion in popularity of the Internet during this time eventually led to its steady penetration into many aspects of life (Leberknight et al., 2012). Between 1993 and 1995, Internet service providers (ISPs) began offering access in most major U.S. cities (Scott & Jacka, 2011). America Online (AOL) became a popular service by mailing access discs directly to consumers, which allowed them to try (and subscribe to) the Internet. Once on the Internet, users could participate in social activities such as sharing their opinions broadly through individual homepages, participating in bulletin board discussions, and engaging in other activities. Widespread use of social media and web 2.0, though, did not proliferate until the 2000s. With the launch of MySpace in 2003, users had the means to control and share media easily on their own personal pages and comment on the contents of their contacts' pages. Scott and Jacka (2011, page 14) argue that "if 2000–2004 was about building platforms and tools, 2005–2009 could be defined as the period of user adoption and the remarkable change in how users connect, converse, and build relationships." By 2006, YouTube and Twitter had launched, and Facebook cracked the 10-million user mark.

Since 2006, the function of the web has moved rapidly in the direction of user-driven technologies such as blogs, social networks, and video-sharing platforms (Smith, 2009). This user-generated content is becoming more prevalent across the web, with most sites now allowing users to publish opinions, share content, and connect with other users.

SOCIAL NETWORKING SITES AND PLATFORMS

Social media use has skyrocketed in the last several years. Participation in social networking sites ballooned from 5% of all adults in 2005 to 50% in 2011. Facebook, in particular, grew from 5 million users in 2005 to more than 900 million in 2012. Twitter continues its amazing growth with nearly 500,000 new accounts created per day (Health Research Institute, 2012).

Web 2.0 tools and technologies in particular have allowed collaboration and communication across boundaries (Schutte, 2009). At the core of web 2.0 are social media sites like Facebook, Twitter,

LinkedIn, YouTube, and many others facilitating social sharing to an unprecedented extent. These sites can be defined as *social network sites* or *social network services*—"web-based services that allow individuals to (1) construct a public or semi-public profile within a bounded system, (2) articulate a list of other users with whom they share a connection, and (3) view and traverse their list of connections and those made by others within the system" (boyd & Ellison, 2007, page 211). On social network sites, individual users share knowledge, pose and solve problems, seek and offer advice, tell stories, and debate issues of interest (Toral et al., 2009). boyd and Ellison differentiate social network sites from *networking* as the latter term implies that relationships are being initiated between strangers. Social networking sites make networking possible, but it is not their primary function nor is it what makes them distinct from other forms of computer-mediated communication. The defining quality of social network sites, according to boyd and Ellison, page 211, is that they "enable users to articulate and make visible their social networks."

Social networking sites host online communities of people who share interests and provide ways for users to interact, including e-mail and instant messaging services (Shin, 2010). Social network sites allow users to create profile pages with personal information, establish "friends" or contacts, and communicate with other users (boyd & Ellison, 2007). On popular sites like Facebook, users communicate via private messages or public comments posted on profile pages. Users also have the option to communicate via instant message, voice, or video. Early adopters of the sites tended to be young people, but an increasing proportion is older. And, indeed, extroverts may use social network sites more often (Sheldon, 2008) as these sites contain relatively public conversations between friends and contacts (Thelwall et al., 2010).

In the United States, 81% of adults go online and 67% of online adults visit social networking sites. Forty percent of those who own a cell phone use a social networking site on their phones (Pew Internet & American Life Project, 2013). Among teens aged 12 to 17, 93% are online and 74% have a profile on a social networking site; social networking is the online activity on which they spend the most time daily (Pew Internet & American Life Project, 2011; Kaiser Family Foundation, 2010). Popular social networking sites and platforms take many different forms, as described in the next section.

Blogs

Blogs (or web logs) are periodically updated sites managed by an individual or group to provide information or opinion on a range of topics. Entries, or posts, are organized so that the most recent entry appears at the top. Many blogs allow and invite discussion from readers in the form of comments directly on the site. In popular blogs, the discussion via comments can contain more varied information and opinion than the original blog post itself. Blogs vary in length, frequency of post, topic, and formality. Because of the advent of free and easy-to-use web publishing tools (e.g., Blogger, WordPress, and Tumblr), blogs are available and used as an outlet for those with a wide range of resources and proficiency with computing. More than 150 million public blogs exist on the web (NielsenWire, 2012). Compared with the general population, bloggers are more likely to be female, in the 18- to 34-year-old age group, well-educated, and active across social media.

Blogs are often interconnected, and writers may read and link to other blogs, providing links to others in their entries. Often referred to as the blogosphere, interconnected blogs are social media in their own right, with social communities and cultures of their own (Mazur, 2010).

Twitter

Twitter has been defined as a "microblogging" service. Launched in 2006, Twitter users or "Tweeters" publish short messages, or "Tweets," up to 140 characters in length. Tweets are publicly visible through the website or third-party applications in real time with millions of Tweets posted across the world per hour. Tweets contain status update information, sharing of links and information, direct messages to other Tweeters, and opinions on almost any topic imaginable (Tumasjan et al., 2010).

With the rise of text-based social media like Twitter and millions of people broadcasting thoughts and opinions on a variety of topics, researchers have taken notice. Twitter, in particular, is convenient for research because of the volume of publicly available messages and the process of obtaining them is relatively simple (O'Connor et al., 2010).

Survey research applications for Twitter are described in Chapters 2, 3, and 9.

Facebook

Launched in 2004, Facebook is a website and social networking service with almost 1 billion active users as of 2012, making it one of the most popular sites on the Internet. Enrollment is free, and users can create a profile and share information such as hometown, current city, education, employment, interests, and favorites. Users can also post photos, videos, notes, and status updates to share with their Facebook "friends." Through Facebook, a user can make new contacts and follow, or "like" different groups, organizations, or products. A survey of 1,487 adults in the United States found that 43% of U.S. adults use Facebook; use is highest among young adults, college graduates, and those with an income greater than $90,000 (Morales, 2011).

Facebook users and their "friends" comprise virtual communities linked by shared interests or opinions. Researchers can sample populations of interest working through existing groups or creating new ones for the purposes of their analysis (Bhutta, 2012). As the number of Facebook users worldwide continues to grow, survey researchers are interested in how we might leverage social media to connect individuals and share information with one another (Lai & Skvoretz, 2011). Potential survey applications using Facebook are described in more detail in Chapter 4.

LinkedIn

Often considered the business world's version of Facebook, LinkedIn is a social networking service for professionals. With more than 150 million users in over 200 countries and territories, LinkedIn provides a means for individuals in various professional fields to interact and discuss issues, build their personal networks, and promote their expertise when seeking new opportunities or employment. Groups serve as hubs of interaction with sharing of announcements, links, and opinions. Over 47 million unique users visit LinkedIn monthly with more than 21 million from the United States.

Second Life

Second Life is an online three-dimensional world in which users (called "residents" in Second Life) design personal avatars and interact with other avatars and their surrounding environment. Communication in Second Life can be through instant messages or voice chat. As opposed to other social media sites like Facebook and Twitter, which typically augment real-life personas and relationships, residents in Second Life represent themselves in ways that depart from real-life appearances and personalities.

Second Life residents come from more than 100 countries. Second Life use is measured in user-hours—and 481 million user-hours were logged in 2009 (Linden, 2011). The most active users (as of 2008) were 25–44 years old (64% of hours logged) and male (59% of hours logged). Second Life provides a context-rich environment for conducting cognitive interviews and other survey pretesting activities (Dean et al., 2009; Murphy et al., 2010). The system allows the researcher to target and recruit specific types of residents through classified-type advertisements, online bulletin boards, and word-of-mouth in the virtual world, which can be more efficient and cost-effective when compared with traditional newspaper ads or flyers that are typically used to recruit in-person cognitive interview subjects (Dean et al., 2011). Text-based chat and voice chat can be collected for analysis, offering full transcripts of cognitive interviews (see Chapter 5). The only elements of in-person cognitive interviews missing from the paradata are facial and physical expressions, although Second Life residents can manipulate these for their avatars to a certain extent. Second Life is also useful in accessing hard-to-reach populations such as those with chronic illnesses (see Chapter 10) and for conducting methodological research (see Chapter 6).

Other Social Networking Platforms and Functionalities

A multitude of other social media platforms and functionalities exist, and the list grows continually. YouTube (owned by Google), for example, is a website for sharing videos. Users can upload original content (e.g., video blogs and performances) or media clips and discuss content on the site using comments. Flickr, Snapchat, and Instagram are popular sites with a similar design but primarily for posting and sharing photographs.

Another social networking site, Foursquare, focuses on physical locations. Users "check in" with their global positioning system

(GPS)-enabled smartphones at various locations and earn awards in the form of points and badges.

For the purposes of this volume, we consider text messaging and smartphone applications (i.e., mobile apps) (although not technically web-based) to fall under the umbrella of social media. Text messaging can be used to interact with study respondents to capitalize on the cultural behaviors of the younger population. A text message is a short communication between mobile phones on a bandwidth lower than that of a phone call, and messages are usually limited to 160 characters (Cole-Lewis & Kershaw, 2010). Mobile apps are standalone pieces of software that a user can download to his or her device that allow for communication and interaction in many ways similar to web-based social media platforms. Chapter 7 discusses mobile apps in more detail.

WHY SHOULD SURVEY RESEARCHERS BE INTERESTED IN SOCIAL MEDIA?

The Current State of Survey Research

Survey research, as a discipline, has been under attack for the better part of several years now because of several coincidental trends that imperil the ability of a survey to provide data that are fully "fit for use"— accurate, timely, and accessible (Biemer & Lyberg, 2003, p. 13). One could argue that, in this day and age of advancing technology, survey data can be made *timelier* and *more* accessible quite easily—but if those data are not accurate, they are not of high quality and are not "fit for use."

These coincidental developments and trends have resulted in an uneasy state of affairs for survey researchers. We now confront declining data quality as a result of, among other issues, falling response rates and inadequate sampling frames.

Falling Response Rates

Virtually everyone in the survey research industry knows that, regardless of mode, survey response rates (the number of completed interviews divided by the eligible sample population) are in decline. Scholars in the survey research field have been decrying the decline since at least the 1990s (de Heer, 1999; Steeh et al., 2001; Tortora, 2004; Curtin et al., 2005). That outcry continues to this day: In his presidential address delivered to members of the American Association for Public Opinion Research (AAPOR) in May 2012, Scott Keeter called the problem of declining response rates "formidable" and noted that they are "well below the norms we were trained to expect." Furthermore, he cited response rates in the single digits in the public opinion (polling) sector of the industry even when conservatively calculated (Keeter, 2012, page 601).

Why are falling response rates a problem or threat to traditional survey research? Data quality has long been a cornerstone of social science research: Researchers are constantly investigating methods to maximize data quality and minimize survey error, which has many components and causes. The failure to obtain data from all sample members, referred to as unit nonresponse, can lead to bias in a researcher's estimates and, thus, flawed conclusions resulting from analysis of survey data if systematic differences exist in the key survey outcomes between responders and nonresponders (Peytchev, 2012). If, for example, the underlying cause for increased unit nonresponse is the *topic* of the survey (e.g., many sampled respondents refuse to participate in a health survey because they view their own health as a sensitive topic), the level of unit nonresponse (the response rate) will almost certainly be correlated with key survey estimates (Groves et al., 2006).

Although researchers and analysts can assess nonresponse bias and its impact in multiple ways, the response rate is the most well-known and most frequently used indicator of error stemming from unit nonresponse. It has been, and remains, "an overall indicator of data quality—interpreted by many as a shorthand way for labeling whether a particular survey is "good," "scientific," "valid"—or not." (Carley-Baxter et al., 2006, page 2).

A combination of factors has likely led to survey researchers' increasing inability to achieve adequate response rates, including increased suspicion regarding requests for data from both government and corporations, increased reluctance to share personal data with unknown persons or entities, and tightened controlled-access to housing units, among others.

Some of the mistrust by potential respondents is no doubt because of burgeoning bids for their attention (and answers/data) from every conceivable angle: The amount of junk mail (both print and electronic) is rising. Telemarketers discovered, early on, the power and ease of calling people at home to ply their wares—so much so, that Americans were inundated with calls that were indiscernible at first from survey requests (Tourangeau, 2004). Some people might have sought to escape the onslaught by declining to answer the door or phone, and instead seeking refuge in their computer, but then spam e-mail and phishing attacks became more prevalent than ever (Kim et al., 2010; Tynan, 2002). The threat of computer viruses from unknown sources, and news stories about identity theft and stolen laptops containing confidential information likely led people to become more protective of their personal information, including refusing to cooperate with a survey solicitation— not bothering to make the distinction between a legitimate survey and any other entreaty for personal information.

People now have all sorts of ways by which they can avoid unwanted solicitations, including requests to participate in surveys. Mobile phones come equipped with caller ID. If people do not recognize the number, they can easily reject the incoming call. Such technology has certainly had an impact on surveys, contributing to declines in response rates and increasing the costs of conducting telephone surveys (Kempf & Remington, 2007; O'Connell, 2010).

Challenges with nonresponse have not been limited to telephone surveys. In-person (or face-to-face or field) surveys have been the gold standard for survey research since at least the 1940s, and most of the massive data collection efforts required by the U.S. federal statistical system are conducted this way (Yeager et al., 2011). But even gold standard surveys are experiencing downward-trending response rates. One reason is the increase of controlled-access housing developments, such as gated communities and buzzer systems in multiunit housing, making it much more difficult for even the most persistent interviewer to contact potential respondents (Keesling, 2008). In recent years, government-funded studies have documented an increase in the amount of nonresponse attributed to controlled access (Cunningham et al., 2005; Best & Radcliff, 2005). Because individuals have been able to restrict access in these ways, reaching them and collecting quality survey data has become more difficult and more costly (Curtin et al., 2005).

Frame Coverage Errors

A second major potential source of error in survey research—and, thus, another nail in the coffin being built by its detractors—is the ever-increasing number of cell-phone-only households and individuals.

Surveys conducted by landline telephone, using a random-digit-dial (RDD) sampling frame, run the risk of missing entire segments of the population of interest if the traditional landline-based telephone sampling methods are not combined with a cell-phone frame (Blumberg & Luke, 2009). The latest data from the U.S. federal government show that more than a third (36%) of American adults now has a cell phone but no landline (see http://www.cdc.gov/nchs/data/nhis/earlyrelease/wireless201212.pdf or Blumberg et al., 2012). Some subgroups are more likely to fall in the cell-phone-only category: adults aged 25–29 (58.1%) and renters (52.5%), for example. These holes in sampling frame coverage have led some researchers to suggest that we abandon telephone sampling and telephone surveys entirely (Holbrook et al., 2003) or that the "entire industry needs to be remade" (O'Connell, 2010).

Other scholars are trying to develop alternative schemes for sampling, acknowledging that something new must be done (see Brick et al., 2011; Link et al., 2008). In this book, we suggest that many social science researchers are now moving beyond alternative sampling schemes and are instead exploring entirely new models of social science data acquisition and analysis. At least some of this search for new ways to find and interpret data is driven by the ready accessibility of data itself: The era of "Big Data" has begun.

The Coming Age of Ubiquity

In Alvin Toffler's *The Third Wave* (1980), he writes that there have been three waves of civilization, beginning with the agricultural wave, followed by the industrialization wave, and now supplanted by the information wave. Toffler further argues that the advent of each wave brings with it uncertainty, upheaval, and change. Those willing to surf along on top or ahead of the new wave—that is, embrace the change—can not only acquire power and wealth, but also fashion fundamental alterations of civilization.

The ubiquity of computers and technology has been the stuff of science fiction for decades. This ubiquity seems tantalizingly close now, but they are still just out of reach. The day is coming when your refrigerator "knows" that it is out of milk and signals the local grocery to add a gallon of 2% milk to your next order. We will not, however, have achieved this state of ubiquity until we no longer take notice of the role that computers and technology play in our everyday lives; that is, implicit in the notion of "ubiquity" is the idea that computers will perform these functions as a matter of course and we will have accepted that.

Variety. Transactions and interactions that are heavily computer dependent already occur on a daily basis. Making a phone call, sending a text message (SMS), making a bank deposit, swiping a loyalty card at Starbucks or Walmart, paying for a meal with a credit card, passing in front of a CCTV camera in London, posting a video on YouTube, querying Google, Tweeting on Twitter, and a hundred other seemingly innocuous activities that we all perform every day all involve a vast network of computers humming away behind the scenes, generating vast amounts of data. All of these bytes of information posted or transacted each day, taken together, equal "Big Data."

Volume. As computers and computer-based technology near ubiquity, the amount of transactional and interactional data created grows exponentially. "In 2012, every day 2.5 quintillion bytes of data (1 followed by 18 zeros) are created, with 90% of the world's data created in the last two years alone," according to one estimate (Conner, 2012, paragraph 3). In February 2011, Facebook users uploaded, on average, 83,000,000 photos *every day*. Walmart generates 1 million customer transactions every *hour*. Every *minute* (http://mashable.com/2012/06/22/data-created-every-minute/), for example, there are:

- 47,000 app downloads from the Apple App Store
- 3,600 pictures uploaded to Instagram
- 27,778 new posts on Tumblr
- 204,166,667 e-mails sent

Velocity. Computers operate at an ever-increasing pace, contributing to the generation of more and more data. Any particular set of variables of interest may now be analyzed on a daily basis, or even hourly or by the second, if preferred. Faster data processing and faster data transmission rates make for shorter and shorter feedback loops, making data and data analysis timelier than ever before.

Validity. However, we know that social media has produced much superfluous material as well. People create Facebook pages for their cats, sensors capture false positives, and data elements go missing. One of the biggest roles that social science researchers can play in this arena is to attempt to make sense of it all: Revolutions in science have often been preceded by revolutions in measurement, and the explosion of data and data accessibility demands new ways of measuring human behavior (Cukier, 2010).

There is a distinction to be made between the "organic" (Groves, 2011) Big Data that are created at a volume of billions of gigabytes per day and the "designed" data that survey researchers have been collecting for nearly a century: Survey methodology, as a field, has focused on developing less costly and more accurate methods of asking people questions so as to obtain self-reports of behaviors and opinions. Those answers populate designed datasets, produced to answer a very specific set of research questions. Those "designed" data are case-poor and are observation-rich in comparison with Big Data; in other words, to control costs, we sample the population of interest and ask a specific set of questions, which results in a dataset with (relatively) few cases, but with just the desired data elements for each case. Organic data, on the other hand, emerging out of new communication technologies and without any specific research purpose or design, grow organically and without boundaries. Separate pieces of data, collected for different reasons and by different entities, can be combined to make even bigger datasets that have the capability to answer questions that neither original dataset could. The result is billions of "cases," but for researchers, the data elements are ill-defined. Nonetheless, this boundless—and boundary-less—set of data has too much potential to ignore: The temptation to use these data to understand social behavior and trends is nearly irresistible. The promise and potential of Big Data, of course, are that they can answer Big Questions. We are on the cusp of an era, a wave, where data are cheap and abundant instead of, as they are now, scarce and expensive to obtain; in fact, a considerable amount of Big Data is actually free to obtain (or, at the very least, so inexpensive as to be an insignificant cost). As the amount of cheap or free Big Data increases, its accessibility (one of the components of data quality, as described above) also goes up, driving its overall value ever higher. One fundamental change that results from these easily accessible and plentiful data is that a business model based on "selling" data will not be sustainable; instead, companies will have to adjust their business model so that what they are selling is analyzed and interpreted data—not the raw data alone.

Perhaps another fundamental change is that, to add value to a set of data, many social science researchers will have to change the way they approach a dataset. Traditionally, survey researchers, statisticians, and data analysts have been accustomed to working with a (relatively speaking) small dataset to which a complex model is applied. With Big Data, we will see more unstructured models applied to huge datasets.

Instead of asking questions of a sampled population, as survey researchers have done for decades, we may now be asking questions of data—instead of people—that already exist. The goal for survey researchers, statisticians, and data scientists as they face the challenge of understanding, analyzing, and drawing conclusions from Big Data will be the same as it has always been for survey researchers: "producing resources with efficient information-to-data ratio" (Groves, 2011, page 869).

Still, at least for the immediate future, interpreting the results of these unstructured models will involve art as well as science. Already, the entire fitness-for-use dynamic is changing: Data, now coming from (or in combination with) "Big," organic sources, instead of just survey (or designed) data, are timelier and more accessible than ever. Social science researchers and survey researchers have a unique opportunity to define and develop the "accuracy" component of the fitness-for-use definition of quality.

Public vs. Private Data

Most Big Data are relatively inexpensive, but there are still questions to be answered about how accessible these data will be. Do we, as researchers, have unfettered and unlimited access to all these Big Data? Although we think that such a complicated issue does not lend itself to an easy answer, others think there *is* an easy answer: "You have zero privacy anyway," said Scott McNealy, the CEO of Sun Microsystems. "Get over it" (*Wired*, 1999, paragraph 2).

Facebook, for example, has come under fire for repeatedly changing its privacy policies (Terms of Use); as of this writing, the policy has 115 clauses, 25 subclauses, 10 appendices, and 13 links to other documents/policies (Facebook, 2012). Most Facebook users likely do not fully and completely understand who can see their latest status update or check-in.

Many companies sell products or services now with the full expectation that the data generated by the product will belong to the company. Nike+ FuelBand and Fitbit, for example, provide users with data about calories burned during exercise, helping users to track their activity level and weight. The device also provides that data to the manufacturer. Other companies (e.g., MC10 and Proteus) make products that monitor heart rate, body temperature, and hydration level. Companies see the value in these (anonymized) data that, once captured, will be quite interesting to healthcare providers (Hardy, 2013).

Social networking sites make public many personal pieces of information about users. Profiles may contain users' name, age, gender, geographic location, and information about hobbies, interests, attitudes, and opinions (Wilkinson & Thelwall, 2011). Most sites have privacy settings that allow users to control who can access their information, but many users leave their profiles public, either by choice or because they do not take the time to understand the distinction, which allows anyone to access this information. As Wilkinson and Thelwall (2011, page 388) explain, "In theory, this gives academics and market researchers unparalleled access to mass personal data with which topics like homophily (Thelwall, 2009; Yuan & Gay, 2006), depression (Goh & Huang, 2009), brand perception (Jansen et al., 2009), and personal taste (Liu, 2007; Liu et al., 2006) can be investigated relatively cheaply and easily."

Social networking sites and platforms allow mass-scale sharing of thoughts, opinions, and behaviors by people worldwide. As a result, a potential goldmine of data is easily accessible, and these data could be harvested to provide insights in the manner that survey data have traditionally been employed.

SOCIAL MEDIA INTERACTION: NEXT WAVE (OR SUBWAVE)?

All of the aforementioned developments suggest that we are approaching the next phase of evolution of survey research. Survey research modes emerge over time as communication technologies and preferences change. Up to this point, the dominant survey research modes have been face-to-face, mail, telephone, and web. Each point along this evolutionary pathway has spawned a new body of research and new methods for increasing the reliability and validity of survey data, thereby reducing the error in the data collected, and increasing its fitness-for-use (quality). As survey researchers with a keen interest in how communication technology shapes the way we ask questions and record answers, we want to know the answers to these questions: Are social media and mobile technology the next wave in survey research? Are these changes a subwave of Toffler's Third Wave? And if they are the next wave or subwave, is the survey research community ready for it?

The history, or evolution, of survey research shows several times at which the research community decided collectively that new modes were acceptable. Early on, survey sampling proponents took advantage of the convergence of (1) well-defined area probability sampling ingredients like census tract information, (2) trends in contemporary print media favoring sharing of individuals' opinions as part of news stories, and (3) an emergent cognitive science that provided tools (questionnaires and scales) for measuring individuals' behaviors and attitudes. These three factors were combined with the communication tools readily available at the time (mail and in-person interaction) to develop the survey interview protocol as we know it (Groves, 2011).

But, the 1960s introduced a departure from that evolutionary vector. Overall, by 1958, 73% of the U.S. population had telephones (Mendelson, 1960), so survey researchers began to consider the telephone as a possible new mode. In 1960, *Public Opinion Quarterly* published a summary of a roundtable discussion from the 15th meeting of the American Association of Public Opinion Research (AAPOR), *Current Status of Telephone Surveys*. In this discussion, presenters pointed out concerns with telephone coverage among less affluent populations, non-white households, and rural areas. Then, in a 1964 study of white women of child-bearing age in the Detroit area, 91.3% of respondents were able to be interviewed by telephone rather than by face-to-face or mail methods (Coombs & Freedman, 1964), helping to convince the research community that the telephone was a viable, valid mode. Coverage grew from 81% in 1963 to nearly 95% in 1983 (Thornberry & Massey, 1988). By eliminating the time and costs of travel for face-to-face surveys and reducing the lag time of mailed surveys, telephone as a mode of data collection increased the efficiency with which survey data could be collected, making survey data both more accessible and more timely.

In the 1990s, another possible departure from the evolutionary pathway took shape. The Pew Research Center for People and the Press first began tracking Internet adoption in June 1995, at which time 14% of adults had access (Pew Internet & American Life Project, 2012). In 2000, Don Dillman (2000) published *Mail and Internet Surveys: The Tailored Design Method* and Mick Couper (2000) published the *Public Opinion Quarterly* piece, "Web Surveys: A Review of Issues and Approaches," which together signaled the beginning of the era of web survey design. By February 2001, Pew reported that 53% of American adults had access to the Internet, and now, Internet penetration in the United States is currently at 81% of adults (Pew Internet & American

Life Project, 2012). The Internet/web is now used readily as a data collection mode, but coverage problems (the "digital divide"), like those of the telephone in the 1950s and 1960s, have prevented it from completely replacing current modes. Instead, Internet data collection is now part of the multimode arsenal or approach adopted by most survey research practitioners to combat the response rate/nonresponse bias problem.

Nevertheless, Internet penetration—and usage among the connected continues to grow. Our definition of social media encompasses a broad range of communication technologies: Social networking sites predominate, but virtual worlds, mobile technologies, blogs and microblogs, Internet telephony services like Skype, and crowdsourcing platforms all fall under the umbrella of our definition. As shown in Figure 1.2, 69% of adult Internet users spend time on social networking sites like Facebook, LinkedIn, or Google Plus; 71% watch videos on sites like YouTube or Vimeo; 53% search online classified ads sites like Craigslist; 46% upload photos; 25% use an online calling service like Skype; 16% use Twitter; 14% author blogs; 8% use online dating services; and 4% use a virtual world like Second Life.

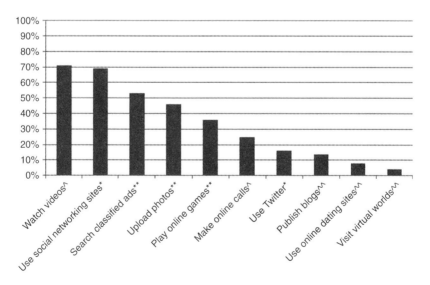

FIGURE 1.2 Percentage of online U.S. population engaging in Internet activities. ^Data last collected in 2011. *Data last collected in 2012. **Data last collected in 2010. ^^Data last collected in 2009. *Source:* Pew: http://pewinternet.org/Trend-Data-(Adults)/Online-Activites-Total.aspx.

Are these rates of adoption and use sufficiently high to be considered the next (sub)wave? Do social media now represent a new mode for survey research? All of these activities generate Big Data, which as noted, results in timely and accessible data. It is time to consider adding methods of capturing and analyzing these data to the survey researcher's toolbox.

ADDING SOCIAL MEDIA TO THE SURVEY RESEARCH TOOLBOX

Already, a plethora of approaches exist that use social media as an adjunct to, or even a replacement for, survey research. Not all of these approaches possess methodological rigor, and certainly, not all find favor with full-probability sampling adherents. Suffice it to say, many techniques are now available for harnessing social media content or social media communication methods as a way of adding to our understanding of human social behavior. Blog posts, Facebook status updates, YouTube comments, online dating or professional profiles, and Tweets all provide passive content (Big Data) about individuals that can be used to supplement our understanding of public opinion. These types of data are passive because researchers do not actively collect them; rather, the data are extant, provided by users, and are available to be analyzed.

For example, researchers can use multiple methods to monitor and mine social media content. Chapters 2 and 3 include two excellent examples of analysis of publicly available social media data, using sentiment analysis, a subfield undergoing rapid growth. Sentiment analysis assigns positive, negative, or neutral codes to text, enabling conclusions about the public sentiment on a particular topic or issue. Researchers can also apply content analysis to passively collected data, going a step beyond sentiment ratings and, instead, looking to uncover recurrent themes in the data. For example, a content analysis of a Twitter conversation about concussions found that 37% of concussion-related content was about recent news stories on concussions, 27% involved a user sharing a personal story or experience relating to concussions, and 13% of content included recommendations for or descriptions of treatment procedures (Sullivan et al., 2011).

Another method of online sentiment and content analysis is called netnography (Kozinets, 2010): ethnographic research that takes place in online communities. Netnographers may use content or sentiment analysis to understand the communities, but as an ethnographic method, it involves some interaction with research subjects; the researcher becomes a participant in the online community to better understand its culture and content. Both content and sentiment analysis may be used to analyze the results of netnographic data collection.

Netnography bridges the gap between passive data collection and the investigator-initiated, more traditional way of eliciting data from participants. And myriad other opportunities exist for data collection from social media platforms beyond immersive cultural studies. In the chapters that follow, survey samples are developed through a mobile panel recruited from iTunes and Google Play, via two crowdsourcing platforms, and from online ads placed on Craigslist and Facebook, from using hashtags on Twitter, and within Second Life user forums. Questions are asked in 140-character Tweets, by avatars inside an immersive three-dimensional (3-D) virtual world, within a Facebook app, within a mobile survey app, as tasks assigned through a crowdsourced company, and as part of games designed to make the process of answering questions more fun.

All of these approaches can be legitimate methods of producing quality data that are timely, accessible, and accurate—although, to be sure, many issues are still to be debated with these new and evolving methods. One way of sorting through the choice of which method to use is to think carefully about the nature of the data to be analyzed. Were these data arrived at passively, or were they produced by direct questioning? Were these data broadcast to a large and impersonal audience, or were they part of one-to-one communication? Did these data come from an online community of like-minded individuals who, collectively, represent some sort of group? Sociality can be a useful concept for cataloguing these data.

TOWARD USING THE CONCEPT OF SOCIALITY IN SURVEY RESEARCH OF THE FUTURE

Sociality can be defined as the extent to which an individual is social or sociable, or the tendency to associate with groups (*The Free Dictionary*). Put another way, the extent to which a person socializes, or interacts with others, is *sociality*. People are inherently social, and they associate with others and form groups for the simple goal of forming social relationships (Fiske, 1992). Thus, human sociality involves establishing identity, organizing social behavior, and convening and communing with other people; it also includes various levels of relationships, including private, public, and family (LaMendola, 2010). Survey researchers recognize that data collection takes place in a social context and, thus, takes into account sociality, whether expressly acknowledged or not. To be sure, sociality can be observed in many stages of the survey interview. Sociality can affect surveys in the following ways:

- Different experiences among respondents' family and relationship dynamics might result in different interpretation of questions about household income.
- Field interviewers observe neighborhood socioeconomic status, coming-and-going times, and friendliness/openness when deciding the best way to approach households selected for data collection.
- Survey designers take steps to establish privacy in the interview setting (self-administration, no identifying names, and private rooms) before posing sensitive questions to respondents.
- Asking questions about sexual behavior in the context of a survey about drug abuse and other risky behaviors might result in different answers than when asked in a survey about family and marital relationships.

The growth of online channels of communication and the resultant social media interaction presage an expansion or alteration of the social context, or sociality, of survey research. If human sociality provides the context in which survey research takes place, then the growth of online sociality provides an entirely new context with new ramifications for survey research.

For example, many research questions in surveys concern private behavior. The next subwave (of the Information Wave) has the potential to alter radically the sociality of a survey research interview. If people are now willing to disclose more private information, as they do on social media sites, they may be more willing to disclose it to researchers, especially if we adopt (or co-opt) new tools for eliciting that information. As multiple examples in this book demonstrate, people are quite willing, using various social media channels, to broadcast information, thoughts, motivations, and ideas that were previously considered private.

Human sociality is both restricted and enhanced by online communication technologies. On the one hand, these new technologies or communication platforms facilitate the construction of identity (profiles), manifest relationships (friend networks/lists), and enable community (within apps, Facebook groups, blogs, and message boards). On the other hand, online social spaces present quite real challenges to individual privacy and boundaries. In an environment so open and oriented toward sharing, it becomes difficult to manage an appropriate level of disclosure for all different types of human relationships (personal, professional, public, family, romantic, and friendly) (Leenes, 2010).

Social network services (SNSs) like Facebook specifically encourage informal forms of sociality including chat, expression of tastes and preferences ("likes"), and gossip. As communication tools, SNSs turn previously informal or chat-type discourse into a formalized record with an audit trail; ultimately, certainly, such trails will change the norms of social behavior/sociality. Data points previously kept private, such as relationship status or sexual orientation, are now aired in public via SNSs and other online profiles. Although SNS users can choose what to disclose, and they can customize privacy settings, any post, entry, or status update has the potential to be disclosed when an SNS's privacy policy is changed (van Dijck, 2012). Despite these privacy concerns, however, many individuals clearly continue to broadcast their innermost thoughts (no matter how profound or mundane) using SNS platforms.

At another level, the expansiveness of online sociality has the potential to change how survey researchers engage respondents in a "conversation with a purpose." The survey interview has long been defined as such (Bingham & Moore, 1924; Cannell & Kahn, 1968; Schaeffer, 1991), but now, perhaps, the limitations (and advantages) of conversing by a tweet or with an avatar in a virtual world suggest that we may be able to redefine the survey research conversation.

As a final example, the growth in online communities—shared conversational environments and interfaces—portends not only another way for sociality to be expressed but also another way to conduct survey research conversations.

This hierarchical way of conceptualizing individuals in their social contexts is not unique to our application of sociality. Our model maps to the social–ecological model (SEM), a tool developed by the Centers for Disease Control and Prevention (CDC) to develop health intervention and prevention programs to address health concerns at multiple levels (McLeroy et al., 1988). The SEM has four levels: individual, relationship, community, and societal, which are interrelated (Bronfenbrenner, 1994). The individual level includes personal and biological traits that contribute to a particular disease, condition, or behavior. The relationship level includes characteristics of family, peer, and romantic relationships that affect health status. The community level includes the setting and environment in which these relationships and individual behaviors occur (for public health considerations, these environs could include neighborhoods, towns, schools, and work environments, among others). The societal level includes social forces that affect health status, including social norms, public policies, and economic markets (Elastic Thought, 2008). The SEM model has been applied to public health problems as diverse as bullying (Swearer, Napolitano & Espelage, 2011), violence (Dahlberg & Krug, 2002), and child abuse (Belsky, 1980).

We now use this hierarchical way of conceptualizing individuals in their social contexts to help understand the impact of online sociality on communications and survey research. Figure 1.3 depicts the sociality hierarchy, a model that has three levels of online communication: broadcast, conversational, and community-based conversation. Each level has different implications and opportunities for survey research via social media and other new communication technologies.

Broadcast social media occur at the individual level. At this level, online users speak to the crowd or from a virtual soapbox, dispensing information about themselves for anyone to consume. Perhaps the best examples of broadcast behaviors are "Tweeting," sharing information via public profiles and blogging. The essence of *broadcast* social media, as we define it, is one person communicating with many others.

Conversational social media occur at the interpersonal level, that is, between two people, similar to analog or face-to-face communication. At this level, the dominant form of conversation is dialogue or one-to-one communication. Examples of conversational behaviors using new technologies and social media include videoconference calls via Skype, sending a text message to a friend, or having a conversation with another avatar in Second Life.

Community-based conversation via social media occurs when groups communicate with each other and within their membership ranks. Examples include Facebook groups organized around a professional

FIGURE 1.3   The sociality hierarchy.

interest, social gaming, and forming support groups on message boards or in virtual worlds. The key concept at this communication level is *sharing* information. That is, at this level, the conversations are many-to-many—an individual may start the conversation, but many will join in, and many will "see" the conversation.

HOW CAN SURVEY RESEARCHERS USE SOCIAL MEDIA DATA?

This volume is organized by the three levels of our sociality hierarchy. Each of the 10 pilot studies detailed in this book employs broadcast, conversational, or community-based social media or technology as an innovative approach to supplementing or enhancing survey research.

Chapters 2 and 3 examine broadcast-level social media. Twitter users broadcast their 140-character thoughts to an audience of followers—and to anyone who uses a screen-scraping tool to #discover trends and keywords in those Tweets. The organic data produced by screen-scraping can be used for nowcasting, sentiment analysis and various other analysis techniques. In Chapter 2, Haney provides a handbook for sentiment analysis, paying close attention to the pitfalls that await researchers who do not carefully consider their research question. In Chapter 3, Kim et al. perform a case study of Tweets on health-care reform to determine whether such analysis could ever replace opinion polls on the topic.

Chapters 4–7 present examples of use of the conversational properties of social media for survey research. In our model, the best use of the conversational level of the sociality hierarchy involves translating or transferring traditional survey methods to new social media platforms. In Chapter 4, Sage describes using a Facebook app to build a sample of respondents and collect survey data. In Chapter 5, Dean et al. demonstrate how the virtual world Second Life and Skype videoconferencing software can be used to conduct cognitive interviews with a geographically dispersed population. In Chapter 6, Richards and Dean use Second Life as a survey laboratory to test comprehension and compliance with the randomized response technique, a method of increasing the privacy of sensitive questions in surveys. In Chapter 7, Roe et al. describe the processes involved and decisions made in building a mobile survey panel.

Chapters 8–11 exemplify the community level of the sociality hierarchy. Community-based research takes advantage of the social and interactive elements of social media, including instant feedback (for example, sharing results with others in the group) and gamification. In Chapter 8, Keating et al. describe how crowdsourcing techniques can be used to supplement survey research. In Chapter 9, Richards et al. present a method for using Twitter to collect diary data from specific Twitter-friendly communities. In Chapter 10, Haque and Swicegood use extant social networks in Second Life to recruit and interview subjects with chronic medical conditions. Finally, in Chapter 11, Puleston describes methods for gamifying surveys—making them more interactive, interesting, and fun for respondents. In our last chapter, Hill and Dever use the sociality hierarchy to think about ways to improve the survey research of the future.

REFERENCES

ANDERSON, P. (2007). What is Web 2.0? Ideas, technologies and implications for education. *JISC Technology and Standards Watch*. Retrieved from http://www.jisc.ac.uk/publications/reports/2007/twweb2.aspx.

ASBERG, P. (2009). Using social media in brand research. Retrieved from http://www.brandchannel.com/images/papers/433_Social_Media_Final.pdf.

BELSKY, J. (1980). Child maltreatment: An ecological integration. *American Psychologist, 34*(4), 320–335.

BEST, S. J., & RADCLIFF, B. (2005). *Polling America: An encyclopedia of public opinion*. Westport, CT: Greenwood Press.

BHUTTA, C. B. (2012). Not by the book: Facebook as a sampling frame. *Sociological Methods & Research, 41*(1), 57–88.

BIEMER, P. B., & LYBERG, L. A. (2003). *Introduction to survey quality*. New York: Wiley.

BIEMER, P. P., HERGET, D., MORTON, J., & WILLIS, W. G. (2000). The feasibility of monitoring field interview performance using computer audio-recorded interviewing (CARI) (pp. 1068–1073). *Proceedings of the American Statistical Association's Section on Survey Research Methods.*

BINGHAM, W., & MOORE, B. (1924). *How to interview*. New York: Harper & Row.

BLUMBERG, S. J., & LUKE, J. V. (2009). Reevaluating the need for concern regarding noncoverage bias in landline surveys. *American Journal of Public Health, 99*(10), 1806–1810.

BLUMBERG, S. J., LUKE, J. V., GANESH, N., DAVERN, M. E., & BOUDREAUX, M. H. (2012). *Wireless substitution: State-level estimates from the National Health Interview Survey, 2010–2011*. National Center for Health Statistics. Retrieved from http://www.cdc.gov/nchs/data/nhsr/nhsr061.pdf.

BOYD, D. M., & ELLISON, N. B. (2007). Social network sites: Definition, history, and scholarship. *Journal of Computer-Mediated Communication, 13*(1), article 11.

BRICK, J. M., WILLIAMS, D., & MONTAQUILA, J. M. (2011). Address-based sampling for subpopulation surveys. *Public Opinion Quarterly, 75*(3), 409–428.

BRONFENBRENNER, U. (1994). Ecological models of human development. In *International encyclopedia of education*, Vol. 3, 2nd ed. Oxford, UK: Elsevier. Reprinted in Gauvin, M., & Cole, M. (Eds.), *Readings on the development of children,* 2nd ed. (1993, pp. 37–43) New York: Freeman.

CANNELL, C. F., & KAHN, R. (1968). Interviewing. In G. Lindzey & E. Aronson (Eds.), *The handbook of social psychology* (Vol. 2, pp. 526–595). Reading, MA: Addison-Wesley.

CARLEY-BAXTER, L., HILL, C. A., ROE, D., TWIDDY, S. E., & BAXTER, R. (2006). Changes in Response Rate Standards and Reports of Response Rate over the Past Decade. Presented at Second International Conference on Telephone Survey Methodology, Miami, FL, January 2006. Retrieved from http://www.rti.org/pubs/TSM2006_Carl-Bax_Hill_paper.pdf.

Cole-Lewis, H., & Kershaw, T. (2010). Text messaging as a tool for behavior change in disease prevention and management. *Epidemiologic Reviews, 32*(1), 20.

Conner, M. (2012). Data on Big Data. Retrieved from http://marciaconner.com/blog/data-on-big-data/.

Coombs, L., & Freedman, R. (1964). Use of telephone interviews in a longitudinal fertility study. *Public Opinion Quarterly, 28*(1), 112–117.

Couper, M. P. (2000). Review: Web surveys: A review of issues and approaches. *Public Opinion Quarterly, 64*(4), 464–494.

Cukier, K. (2010). Data, data everywhere. *The Economist*. Retrieved from http://www.economist.com/node/15557443.

Cunningham, D., Flicker, L., Murphy, J., Aldworth, W., Myers, S., & Kennet, J. (2005). *Incidence and impact of controlled access situations on nonresponse*. Presented at the annual conference of the American Association for Public Opinion Research, Miami Beach, FL.

Curtin, R., Presser, S., & Singer, E. (2005). Changes in telephone survey nonresponse over the past quarter century. *Public Opinion Quarterly, 69*, 87–98.

Dahlberg, L. L., & Krug, E. G. (2002). Violence-a global public health problem. In E. Krug, L. L. Dahlberg, J. A. Mercy, A. B. Zwi, & R. Lozano (Eds.), *World report on violence and health* (pp. 1–56). Geneva: World Health Organization.

de Heer, W. (1999). International response trends: Results of an international survey. *Journal of Official Statistics, 15*(2), 129–142.

De Souza, C. S., & Preece, J. (2004). A framework for analyzing and understanding online communities. *Interacting with Computers, The Interdisciplinary Journal of Human-Computer Interaction, 16*(3), 579–610.

Dean, E., Cook, S. L., Keating, M. D., & Murphy, J. J. (2009). Does this avatar make me look fat? Obesity & interviewing in Second Life. *Journal of Virtual Worlds Research, 2*(2), 11.

Dean, E., Cook, S., Murphy, J., & Keating, M. (2011). The effectiveness of survey recruitment methods in Second Life. *Social Science Computer Review, 30*(3), 324–338.

Dillman, D. (2000). *Mail and Internet surveys: The tailored design method*. New York: Wiley.

Elastic Thought (2008). Social-ecological model—the "other" SEM. Retrieved from http://elasticthought.com/2008/10/social-ecological-model—-the-other-sem.html.

Facebook (2012). Statement of rights and responsibilities. Retrieved from http://www.facebook.com/legal/terms.

Fiske, A. P. (1992). The four elementary forms of sociality: Framework for a unified theory of social relations. *Psychological Review, 99*(4), 689–723.

Goh, T., & Huang, Y. (2009). Monitoring youth depression risk in Web 2.0. *VINE, 39*(3), 192–202.

Groves, R. M. (2011). Three eras of survey research. *Public Opinion Quarterly, 75*(5), 861–871.

GROVES, R. M., COUPER, M. P., PRESSER, S., SINGER, E., TOURANGEAU, R., ACOSTA, G. P., & NELSON, L. (2006). Experiments in producing nonresponse bias. *Public Opinion Quarterly, 70*(5), 720–736.

HARDY, Q. (2013). Big data in your blood. Retrieved from http://bits.blogs.nytimes .com/2012/09/07/big-data-in-your-blood/.

Health Research Institute (2012). Social media "likes" healthcare: From marketing to social business. Retrieved from http://www.pwc.com/us/en/health-industries/publications/health-care-social-media.jhtml.

HOLBROOK, A. L., GREEN, M. C., & KROSNICK, J. A. (2003). Telephone versus face-to-face interviewing of national probability samples with long questionnaires. *Public Opinion Quarterly, 67*(1), 79–125.

JANSEN, B. J., ZHANG, M., SOBEL, K., & CHOWDURY, A. (2009). Twitter power: Tweets as electronic word of mouth. *Journal of the American Society for Information Science and Technology, 60*(11), 2169–2188.

Kaiser Family Foundation (2010). Daily media use among children and teens up dramatically from five years ago. Retrieved from http://www.kff.org/entmedia/ entmedia012010nr.cfm.

KEESLING, R. (2008). Controlled access. In P. Lavrakas (Ed.), *Encyclopedia of survey research methods.* Thousand Oaks, CA: Sage.

KEETER, S. (2012). Presidential address: Survey research, its new frontiers, and democracy. *Public Opinion Quarterly, 76*(3), 600–608.

KEMPF, A. M., & REMINGTON, P. L. (2007). New challenges for telephone survey research in the twenty-first century. *Annual Review of Public Health, 28*, 113–126.

KIM, W., JEONG, O.-R., KIM, C., & SO, J. (2010). The dark side of the Internet: Attacks, costs and responses. *Information Systems.* Retrieved from http://www.sciencedirect .com/science/article/B6V0G-51J9DS4-1/2/1a75ce9706b13898decd576f47395638.

KOZINETS, R. (2010). *Netnography: Doing ethnographic research online.* Thousand Oaks, CA: Sage.

KRISHNAMURTHY, B., & CORMODE, G. (2008). Key differences between Web 1.0 and Web 2.0. *First Monday, 13*(6).

LAI, J. W., & SKVORETZ, J. (2011). *Utilizing Facebook application for disaster relief: Social network analysis of American Red Cross cause joiners.* Presented at the 2011 American Association for Public Opinion Research, Phoenix, AZ.

LAMENDOLA, W. (2010). Social work and social presence in an online world. *Journal of Technology in Human Services, 28*, 108–119.

LEBERKNIGHT, C., INALTEKIN, H., CHIANG, M., & POOR, H. V. (2012). The evolution of online social networks a tutorial survey. *IEEE Signal Processing Magazine, 29*(2), 41–52.

LEENES, R. (2010). Context is everything: Sociality and privacy in online social network sites. In M. Bezzi et al. (Eds.), *Privacy and identity.* IFIP AICT *320*, 48–65.

LINDEN, T. (2011). *2009 end of year Second Life economy wrap-up (including Q4 economy in detail).* Retrieved from http://community.secondlife.com/t5/Features/

2009-End-of-Year-Second-Life-Economy-Wrap-up-including-Q4/ba-p/653078/ page/3.

Link, M. W., Battaglia, M. P., Frankel, M. R., Osborn, L., & Mokdad, A. H. (2008). A comparison of address-based sampling (ABS) vs. random digit dialing (RDD) for general population surveys. *Public Opinion Quarterly, 72*(1), 6–27.

Liu, H. (2007). Social network profiles as taste performances. *Journal of Computer-Mediated Communication, 13*(1), article 13.

Liu, H., Maes, P., & Davenport, G. (2006). Unraveling the taste fabric of social networks. *International Journal on Semantic Web and Information Systems, 2*(1).

Mazur, E. (2010). Collecting data from social networking Web sites and blogs. In S. D. Gosling & J. A. Johnson (Eds.), *Advanced methods for conducting online behavioral research* (pp. 77–90). Washington, DC: American Psychological Association.

McLeroy, K. R., Bibeau, D., Steckler, A., & Glanz, K. (1988). An ecological perspective on health promotion programs. *Health Education & Behavior, 15*(4), 351–377.

Mendelson, H. (1960). Round-table sessions current status of telephone surveys. *Public Opinion Quarterly, 24*(3), 504–514.

Morales, L. (2011). Google and Facebook users skew young, affluent, and educated. Retrieved from http://www.gallup.com/poll/146159/facebook-google-users-skew-young-affluent-educated.aspx.

Murphy, J., Dean, E., Cook, S., & Keating, M. (2010). *The effect of interviewer image in a virtual-world survey*. RTI Press.

NielsenWire (2012). Buzz in the blogosphere: Millions more bloggers and blog readers. Retrieved from http://blog.nielsen.com/nielsenwire/online_mobile/buzz-in-the-blogosphere-millions-more-bloggers-and-blog-readers/.

O'Connell, A. (2010). Reading the public mind. *Harvard Business Review*. Retrieved from http://hbr.org/2010/10/reading-the-public-mind/ar/pr.

O'Connor, B., Balasubramanyan, R., Routledge, B., & Smith, N. (2010). From Tweets to polls: Linking text sentiment to public opinion time series. *Proceedings of the Fourth International AAAI Conference on Weblogs and Social Media*.

Peytchev, A. (2012). Multiple imputation for unit nonresponse and measurement error. *Public Opinion Quarterly, 76*(2), 214–237.

Pew Internet & American Life Project (2011). Web 2.0. Retrieved from http://www.pewinternet.org/topics/Web-20.aspx?typeFilter=5.

Pew Internet & American Life Project (2012). Retrieved from http://pewinternet.org/Trend-Data-(Adults)/Internet-Adoption.aspx.

Pew Internet & American Life Project (2013). Retrieved from http://pewinternet.org/Pew-Internet-Social-Networking-full-detail.aspx.

Ravenscroft, A. (2009). Social software, Web 2.0 and learning: Status and implication of an evolving paradigm. *Journal of Computer Assisted Learning, 25*(1), 1–5.

SCHAEFFER, N. C. 1991. Conversation with a purpose—or conversation? Interaction in the standardized interview. In P. Biemer, R. Groves, L. Lyberg, N. Mathiowetz, & S. Sudman (Eds.), *Measurement errors in surveys*. New York: Wiley.

SCHUTTE, M. (2009). Social software and Web 2.0 technology trends. *Online Information Review, 33*(6), 1206–1207.

SCOTT, P. R., & JACKA, J. M. (2011). Auditing social media: A governance and risk guide. ISBN: 978-1-118-06175-6. Wiley.

SHELDON, P. (2008). The relationship between unwillingness-to-communicate and Facebook use. *Journal of Media Psychology: Theories, Methods, and Applications, 20*(2), 67–75.

SHIN, D. H. (2010). Analysis of online social networks: A cross-national study. *Online Information Review, 34*(3), 473–495.

SMITH, T. (2009). Conference notes-the social media revolution. *International Journal of Market Research, 51*(4), 559–561.

STEEH, C., KIRGIS, N., CANNON, B. & DEWITT, J. (2001). Are they really as bad as they seem? Nonresponse rates at the end of the twentieth century. *Journal of Official Statistics, 17*(2), 227–247.

SULLIVAN, S. J., SCHNEIDERS, A. G., CHEANG, C., KITTO, E., HOPIN, L., WARD, S., AHMED, O., & MCCRORY, P. (2011). What's happening? A content analysis of concussion-related traffic on Twitter. *British Journal of Sports Medicine* (published Online March 15, 2011).

SWEARER NAPOLITANO, S. M., & ESPELAGE, D. L. (2011). Expanding the social-ecological framework of bullying among youth: Lessons learned from the past and directions for the future [Chapter 1]. *Educational Psychology Papers and Publications*. Paper 140. http://digitalcommons.unl.edu/edpsychpapers/140.

The Free Dictionary (2013). Retrieved from http://www.thefreedictionary .com/sociality.

THELWALL, M. (2009). Introduction to Webometrics: Quantitative web research for the social sciences. *Synthesis Lectures on Information Concepts, Retrieval, and Services, 1*(1), 1–116.

THELWALL, M., WILKINSON, D., & UPPAL, S. (2010). Data mining emotion in social network communication: Gender differences in MySpace. *Journal of the American Society for Information Science and Technology, 61*(1), 190–199.

THORNBERRY, O. T., & MASSEY, J. T. (1988). Trends in United States telephone coverage across & time and subgroups. In R. M. Groves, P. P. Biemer, L. E. Lyberg, J. T. Massey, W. L. Nicholls, & J. Waksberg (Eds.), *Telephone survey methodology* (pp. 25–49). New York: Wiley.

TOFFLER. A. (1980). *The third wave*. New York: Bantam.

TORAL, S. L., MARTÍNEZ-TORRES, M. R., BARRERO, F., & CORTÉS, F. (2009). An empirical study of the driving forces behind online communities, *Internet Research, 19*(4), 378–392.

TORTORA, R. D. (2004). Response trends in a national random digit dial survey. *Metodolski zvezki, 1*(1), 21–32.

TOURANGEAU, R. (2004). Survey research and societal change. *Annual Review of Psychology, 55,* 775–801.

TUMASJAN, A., SPRENGER, T. O., SANDERS, P. G., & WELPE, I. M. (2010). Predicting elections with Twitter: What 140 characters reveal about political sentiment (pp. 178–185). *Proceedings of the Fourth International AAAI Conference on Weblogs and Social Media.*

TYNAN, D. (2002, August). Spam Inc. *PC World.*

VAN DIJCK, J. (2012). Facebook as a tool for producing sociality and connectivity. *Television & New Media, 13,* 160.

WILKINSON, D., & THELWALL, M. (2011). Researching personal information on the public web: Methods and ethics. *Social Science Computer Review, 29*(4), 387–401.

Wired. (1999, January 26). Sun on privacy: 'get over it.' Retrieved from http://www.wired.com/politics/law/news/1999/01/17538.

YEAGER, D. S., KROSNICK, J. A., CHANG, L., JAVITZ, H. S., LEVENDUSKY, M. S., SIMPSER, A., & WANG, R. (2011). Comparing the accuracy of RDD-telephone surveys and Internet surveys conducted with probability and non-probability samples. *Public Opinion Quarterly, 75*(4), 709–747.

YUAN, Y. C., & GAY, G. (2006). Homophily of network ties and bonding and bridging social capital in computer-mediated distributed teams. *Journal of Computer-Mediated Communication, 11*(4), 1062–1084.

CHAPTER 2

Sentiment Analysis: Providing Categorical Insight into Unstructured Textual Data

Carol Haney, *Toluna*

Sentiment, in text, is the appearance of subjectivity. That subjectivity may be an opinion or an emotion the author communicates and the reader then perceives. Both components—author intent and reader perception—are important in determining sentiment in text. An example of opinion is "the weather is mild today," where "mild" is the author's interpretation of the temperature. What is mild to someone in the Arctic zone may be downright chilly to someone who lives near the Equator. On the other hand, objective facts are those independent of the author's interpretations. For example, "the weather is 76 degrees today with 0% humidity" is objective and, as such, has neutral sentiment.

In the online survey and social media domains, a significant amount of digital data is unstructured; that is, these data are considered text. Textual data are difficult to analyze because they do not have a fixed numeric, nominal, or ordinal structure. Textual data, if structured, almost always map to a categorical (or nominal) structure. All textual data, if

Social Media, Sociality, and Survey Research, First Edition.
Edited by Craig A. Hill, Elizabeth Dean, and Joe Murphy.
© 2014 John Wiley & Sons, Inc. Published 2014 by John Wiley & Sons, Inc.

separated into atomic units of an idea, have an associated sentiment, which may be positive, negative, or neutral. This chapter discusses both the conceptual implications of applying sentiment to textual data as well as the operational steps to apply structure to the text under consideration. Digital data, especially online, increasingly include image, video, and audio content. Some techniques used with textual data also apply to these media, and other, new techniques are available for emotion and sentiment detection specific to these media, such as real-time facial coding.

DESCRIBING EMOTIONAL OR SUBJECTIVE FEELING IN TEXTUAL DATA

Text is an unstructured communication of one or more ideas by a writer. The text is considered unstructured because it is not confined to a finite set of categories, such as the selection of the text "apple" from a finite list of fruits (apple, banana, orange, and kiwi, for example). Unstructured, written text flows from the writer's mind through his or her fingertips on a keyboard into a printed form—and this printed form today is almost always digital. The text may be short (a few words) or long (hundreds of sentences that form a document). The length of the specific text is conscribed by a combination of input mode (such as mobile phone or computer), media channel (social media or e-mail), and the author's intent.

Authors today create digitally unstructured text using many different input modes, e.g., mobile phones, smartphones, computers, and tablets, and converting speech into text. Each mode affects the form and voice of the digitized text. For example, when one is typing an SMS (or text) message into a mobile phone with a 12-key keypad, the limitation of using that keypad necessarily impacts the length and words used to convey meaning.

The media channel, or application, where that text is created also affects both meaning and content. Media channels include social media applications, like Facebook or Twitter, e-mail software, blogging software, review sites (such as Amazon or Yelp), or an online survey. The channel's settings of privacy, perceived readership, and preset length of content may affect the author's voice, content, and often, the sentiment shared.

The author's intent may overshadow the strictures that the input mode and communication channel impose on the text's length, tone, purpose, and sentiment. Although a text message usually communicates a single idea in a short message, this format may not be the final structure of the text in its final context. For example, Tracey Moberly, in the book *Text-Me-Up* (2011), weaved together thousands of text messages (either authored or sent to her) from over a decade into a full-length autobiographical memoir. Another example is Leif Peterson's almost daily Facebook posts over 4 months, including comments and interactions with Facebook friends, which were published as the novel *Missing* (2010). Both of these longer texts were edited—however slightly—into more cohesive final forms. Although these examples are extreme expansions of the usual purpose of the communication channels, the extension applies to the social research method—concatenating the separate textual units from a specific individual, or respondent, gives a similar autobiographical insight, which (once structured) enlarges the respondent's understanding in a way that is difficult to do within a directed survey. Qualitative research, such as focus groups and in-depth interviews, strives for similar autobiographical insight, necessarily with fewer participants and more directives.

Opinion sometimes, but not always, may have a discernible, perceived related emotion. An example of sentiment as an emotion perceived in text is "pork rinds repulse me." The emotion discerned is disgust, which is usually regarded as a negative sentiment. Although an opinion may be an unwritten interpretation of objective facts ("the weather is mild and warm today"), no distinct emotion may be perceived from this statement (even though the writer may be a skier who looks forward to cold, snowy weather).

DEFINITION OF MACHINE-AUGMENTED SENTIMENT ANALYSIS

Sentiment analysis is the process of trying to predict sentiment as opinion or emotion using a consistent, repeatable, algorithmic approach when reading the text under analysis. During the execution of sentiment analysis, either a human being or a computer may identify the sentiment in the text. Most likely, human beings and computers (also called machines in some of the terms associated with sentiment analysis and its related field, text analytics or text mining) work together to identify the sentiment. A primary end result of sentiment analysis is to classify within a specific set of representative categories the sentiment of each text under analysis.

Usually, sentiment analysis is applied to a set of multiple texts, identifying the sentiment of each text. This set of multiple texts is often called a dataset or a corpus; each text is often called a verbatim (defined as the individual textual response a respondent gives to an open-ended question). If the data are from social media, the individual text is called a post (in this chapter, each text is called a post, no matter the source). A post is otherwise known as a single comment entered by someone online within a social network or other forum, such as a blog. Texts may vary in length, media channels, authors, and other attributes. The dataset probably contains numerous posts or texts because many techniques for determining sentiment require splitting the dataset into training and test data as well as into actual production data.

A rigorous sentiment analysis process is consistent: The same procedure is used to identify sentiment in all texts under a single analysis. Consistency increases the chances that a single text or post is coded the same way multiple times. A rigorous sentiment analysis process should also be repeatable: The same algorithms that were applied to the first dataset under analysis should also be applied to any subsequent datasets. Any changes to those algorithms should be understood within the boundaries of the analysis findings.

How Sentiment Analysis Is Used with Text Data

Given the possibility that both machines and humans will occasionally misinterpret author intent, one may assume that sentiment analysis is too error-prone to add value to analysis; indeed, this often is the case. But for those situations where textual datasets are very large, even with large error rates, machine-based sentiment analysis (especially coupled with other text analytic techniques) provides a singular way to understand otherwise unexamined data. For example, a retailer of casual clothing continuously collects survey data from customers who shop in the stores. After every shopping experience, customers can answer the survey, so the retailer receives annually almost a million open-ended responses to the question, "Do you have any other comments you would like to share with us about [the retail store]?" Manually coding the verbatim responses may be impractical and too costly. Applying sentiment analysis as well as a clustering technique, even with a significant error rate, allowed the retailer to find emerging issues that other parts of the survey or other monitoring had not identified.

The retailer's research analyst used several text analytics procedures to find these issues. First, a quick filtering of the textual data removed those verbatims that were of little value—"no comments" and their variants. This filtering reduced the dataset about 35%, leaving approximately 600,000 responses, which is still too large of a dataset for manual coding. The textual comments varied in length, although the average response was about 40 words. Next, a sentiment analysis technique (discussed below) was applied, separating each response into one of three categories—positive, neutral, or negative. Neutral and positive responses were set aside, leaving fewer than 200,000 responses classified as negative, which were then further classified using a clustering technique that looked for common themes and concepts in the responses.

At this point, a research analyst manually explored the themes and reviewed the responses that fell under those themes. Natural language processing (NLP) techniques were used for preprocessing the textual data, as well as for supporting the sentiment analysis and clustering techniques. These procedures (plus the manual exploration of the themes) identified multiple issues of which the retailer was unaware. For instance, some consumers expressed that a particular cut of clothing was unattractive. Another issue was substandard fabric that degraded within days after the respondent initially wore the item. Because many respondents expressed opinions (these were not single events), the issues could have negatively impacted the retailer if they had been left unidentified.

A word of caution here: Many responses scored during this sentiment analysis—perhaps 30% or more—were inaccurately coded. This inaccuracy was apparent from a manual review of a sample of the responses. And additional themes were probably not identified that might have negatively affected the retailer. Sentiment analysis is not a silver bullet—it is not 100% accurate and does not identify all issues. Despite these limitations, however, the technique gave the retailer awareness of a previously unknown subset of negative themes.

Exploratory

In addition to analyzing open-ended responses to survey data, social media websites can be "scraped" to generate domains and items for survey research. Internet scraping means programmatically extracting textual data from websites and storing information as a dataset for further analysis. In one case study, a dataset was created by scraping publicly available, social media websites that met a certain filtering criteria: a swimsuit manufacturer brand name, the phrase "swim suit" (and variants), and sites that had a modification date within the past 2 months. This case study included websites from Twitter, Facebook, blog sites, message boards, wikis, YouTube, and others. The scrape returned over 200,000 posts.

The posts then became the texts within the analytic dataset. Similar text analytics procedures as described above were used to filter the dataset. One difference in this dataset was the identification within the posts of text memes, or posts that were repeated hundreds of times by multiple authors. An example text meme that showed up in initial analysis was "*Dear boys, if you don't look like Calvin Klein models, don't expect girls to look like Victoria's Secret angels*"—this text had a high association with the target brand and the phrase "swim suit." These text memes, while entertaining, most likely do not give the new information being sought, so they were removed from consideration using NLP techniques. One might argue that removing these quotes from consideration may bias the conclusions; however, in looking at the purpose of the research—trying to find issues with the brand or the swimsuits—these types of posts did not meet those criteria.

Next, sentiment analysis was applied to filter for negative posts, and then these posts were clustered into potential themes that required additional review. Research analysts identified and reviewed many clusters (and the posts that fell into the clusters) to find potential, newly emerging issues. Not all textual data are important to the goal of exploratory research. The issue has to be verifiable using another research approach. One such issue that met this criterion was a recurring complaint that the brand's retail stores, as opposed to catalogs, may not be carrying the right size in swimwear. To quote one blogger: "[Brand] stores need to carry bikinis in the *bigger* sizes in store" (emphasis added). Given the high error rate that sentiment analysis returns, and the lower number of posts per cluster, the possible issue of unavailable bigger sizes was not able to stand on its own for decision making; yet it provided a basis for further research. A quantitative online study of 18- to 34-year-old females was executed, identifying customers who had entered a swimsuit store in past 6 months and had shopped for the specific brand under review. The screened individuals were then asked whether they had purchased a suit from the specific brand and, if not, why they did not purchase. More than 25% of 18- to 34-year-olds said they were unable to find their size in the swimsuit they wanted from the brand. In this case study, the potential issue identified using sentiment analysis with social media data was then validated using survey research techniques.

Formative Research

Another reason to use sentiment analysis on social media data is to assist in formative research. A consumer education campaign wanted to understand the best ways to reach young adults at risk of developing addiction to tobacco. This campaign wanted to understand the motivators that may influence young adults to stop the addiction. The approach was to scrape Facebook and Twitter conversations, looking at ways that people spoke about tobacco with their peers. This scraping returned a very large textual dataset. Sentiment analysis was applied to the conversations, negative conversations were identified, and several recurring themes were identified. First, death of a loved one potentially caused by tobacco use had strong negative emotion. Equally as strong were discussions of watching loved ones live with chronic illness caused by the addiction. These messages were then validated in quantitative message platform testing, eventually becoming part of the formative education campaign.

Continuous Measurement

Sentiment analysis is also used for continuous measurement of customer feedback. A large technology hardware provider created the ability to use machine-based sentiment analysis of customer feedback from multiple channels to identify negative emotions or opinions generated based on a change in the customer-facing business model. The inputs to this analysis came from online helpdesk notes, interactive voice recognition (IVR) feedback systems, call center notes, call center conversation transcriptions, and open-ended customer satisfaction survey questions.

Annually, the customer feedback system processed more than 6.5 million customer comments with potential sentiment. In this system, the sentiment analysis techniques were enhanced by precursors to analysis. One precursor created—where possible—a tightly controlled problem domain of input that improved sentiment scoring. For example, the IVR system provided words as input that customers could use when giving feedback that the sentiment analysis algorithms had picked up as an appropriate signal to denote specific sentiment scoring. In the same way, the online helpdesk may suggest certain phrasing when the customer is filling out the online form that helps appropriately score the feedback. Another precursor tuned the sentiment engine prior to full-scale operations by using human validation of a certain percentage of texts. In addition, during full production, every aggregated report allowed users to drill into the specific verbatim that made up that report and report back inaccuracies. As corporate incentives were partly based on the customer satisfaction reports, almost every negative verbatim was reviewed.

DIFFERENT WAYS OF REPRESENTING SENTIMENT

The primary result of sentiment analysis is a classification of each post in the dataset using a predefined set of categories. Depending on the unique attributes of the input text and the algorithms used to determine sentiment, different types of ordinal scales are used. Alternatively, coding text into nominal categories that define a specific emotional experience may be in some instances more applicable to the analysis. Finally, some posts or text express no sentiment at all—in this case, the text is coded neutral, and depending on the type of scale, the neutral classification has different representations.

Ordinal Scales

Sentiment scales can be binary or given a degree of polarity. On a binary scale, a post is negative, neutral, or positive; no other degree of sentiment is determined as part of the classification scheme. For example, one algorithm might associate the post's sentiment scores with a range of -5 through $+5$. Any given sentiment score from -5 to -1 is then considered negative, -0.99 to 0.99 is considered neutral, and 1 to 5 is considered positive (see Figure 2.1).

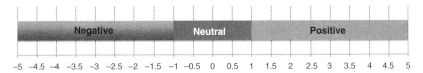

FIGURE 2.1 Binary scale example.

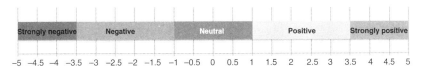

FIGURE 2.2 Degree scale example.

It may seem odd that in this instance, the binary scale has three categories instead of two. Neutral posts do not, in this specific example, participate in the polarity. Note that bands and thresholds are conventionally determined. Any point scale may be used as long as it is consistent and offers polarity. If degrees are desired within each of the sentiment poles (negative/positive), the range may accommodate gradations of polarity. For example, negative sentiment may be classified into strongly negative and negative categories, with similar gradations for positive sentiment. A sentiment score of −4.2 may then be classified as strongly negative if the strongly negative band is −5 to −2.5; again, this is conventionally determined, and in no way should a degree scale be considered more accurate than a binary scale (see Figure 2.2).

Nominal Emotion Classification

Another type of scale is emotional classification. In this method, a numeric score of polarity is not the primary determination of sentiment (although it might be part of the algorithm when determining the emotional category). Instead, with emotional classification, a post is determined to exhibit an emotional quality, such as anger, disgust, fear, joy, sadness, and surprise. Often, emotion classification uses a "bag of words" approach (described in detail in the next section). For example, words such as "furious," "irate," and "pissed off" connote anger, if controlling for negation ("I am *not* furious") and modifiers ("I am not as much irate as I am irritated"). Emotional classification is a form of sentiment analysis that uses a nominal scale instead of an ordinal one. The use of the appropriate scale in sentiment analysis depends on the variability of the subject matter in the input textual dataset, the media channel of the textual data, the input mode, the algorithms used, the accuracy of the scoring, and the content being communicated to consumers of the analysis.

Neutral Sentiment

Neutral posts are cases in which no determinable sentiment is detected; the post may declare a simple fact ("It is raining") or ask a question ("Where are u?"). No matter the scale used to represent the analyzed data, neutral posts almost always make up the largest portion of the input dataset. Posts classified as neutral may have mixed or (perhaps only to the machine performing the classification) conflicting sentiment. Some academic research indicates that existing algorithms tend to score neutral posts as more negative than positive (Moilanen & Pulman, 2007). Additionally, taking neutrally scored posts into consideration may improve overall sentiment accuracy by "stacking" the results of specific algorithms that code into various binary scales and overtly allowing for the specific relationships between negative and neutral posts, and positive and neutral posts.

TECHNIQUES FOR DETERMINING SENTIMENT

The goal for sentiment scoring is to determine as accurate a score as possible to represent the sentiment within the text. Several strategies may improve accuracy when preparing data for analysis and processing the data to determine sentiment. These strategies follow a specific process: first, a precursory exercise of understanding and planning the problem domain; second, harvesting the data; third, structuring and understanding the data; and fourth, analyzing the data (see Figure 2.3).

Precursors to Analysis

In preparing data for analysis, two key strategies are (1) take time to prepare the textual corpus (the set of multiple texts under analysis) to be as complete as possible, and (2) form a tight topic domain. A

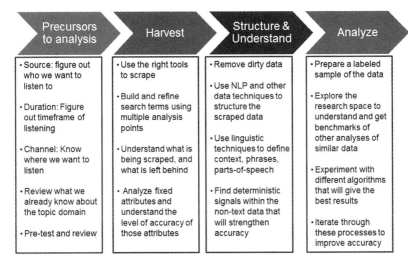

Precursors to analysis	Harvest	Structure & Understand	Analyze
• Source: figure out who we want to listen to • Duration: Figure out timeframe of listening • Channel: Know where we want to listen • Review what we already know about the topic domain • Pre-test and review	• Use the right tools to scrape • Build and refine search terms using multiple analysis points • Understand what is being scraped, and what is left behind • Analyze fixed attributes and understand the level of accuracy of those attributes	• Remove dirty data • Use NLP and other data techniques to structure the scraped data • Use linguistic techniques to define context, phrases, parts-of-speech • Find deterministic signals within the non-text data that will strengthen accuracy	• Prepare a labeled sample of the data • Explore the research space to understand and get benchmarks of other analyses of similar data • Experiment with different algorithms that will give the best results • Iterate through these processes to improve accuracy

FIGURE 2.3   Phases of sentiment analysis process.

well-prepared textual corpus requires the dataset to be as complete as possible. Textual data are usually available to a sentiment analysis process, as well as, attributes about the author of the data, the date, the mode and input of the data entry, and others. These attributes improve the accuracy by participating as inputs into the actual analysis, some of which act as signals or help assist in the determination of sentiment analysis. When online data are scraped, these attributes are available as part of the scraping techniques—even if some data attributes are missing, these attributes should be parsed into the corpus.

Another factor in improving accuracy is to form a tight topic domain of the data under analysis. For example, if one is scraping online data, the goal is to scrape an entire message forum for a specific, particular mention. One should scrape the text for that specific, particular mention— and confirm that it is the intended mention that is being scraped, not a false-positive standing in for that word or mention. If the specific mention was Gap (retail establishment), for example, all posts about "mind the gap" (London Underground saying), "Cumberland Gap," and the like are removed from consideration.

Sometimes, the corpus, when examined prior to harvesting, may contain signals—or information within the data itself—that is highly predictive of a particular sentiment. For example, in Twitter data, emoticons are often used, and as the number of characters of each tweet is limited, an emoticon may become a signal the sentiment algorithm can use. Assigning sentiment based only on signals such as emoticons is not usually a recommended approach for sentiment analysis; yet signals, combined with other techniques, significantly improve accuracy.

Harvesting

In textual analysis techniques, interplay occurs among designing the model, applying it, evaluating the results, and then refining the model to incorporate the findings and results to get a better dataset. The following sections discuss the model, review the type of results expected, and then discuss refinement of the model.

Before any sentiment analysis takes place, textual data must be acquired, including any structured information attached to the textual data; then the entirety must be preprocessed into a textual dataset structured for analysis. This task can be quite extensive and requires an iterative approach. Getting the right textual data is the first step in determining the failure or success of any downstream analysis.

For social media data, data acquisition is often performed by building a dictionary of terms used to match against relevant data taken from the Internet. These search terms limit the text data that are analyzed. (Note: Sometimes the entire site is the topic domain in itself.) The terms are specific phrases that define the topical domain of the research subject.

The search terms should not be too narrow as to limit the analysis nor should they be overly broad. Some data that are not relevant to analysis (called false positives) will become part of the text corpus; the task is to limit the false positives to a small number. Search terms are made up of the phrases common to the topical domain. Often those phrases are used only within a specific number of words from other phrases, or when other words or phrases (also called terms) are not used coterminously. A term can be a word or a phrase. Table 2.1 defines the components that make up search criteria.

The examples here are simplified; in actuality, search terms can be quite complex, with co-occurring words and stop word lists in the hundreds and with statistical calculations determining proximity. Many techniques can be used to build a set of search terms to perform scraping. One technique is to analyze the data for term frequency to identify co-occurring terms within the dataset and, if appropriate, add those new

TABLE 2.1
Example Components of Search Criteria

Type of Search	Definition	Example
Primary search terms	Exact terms found within the candidate page being considered for scraping. In this example, the page should be scraped and added to the corpus if one or more of the terms "marijuana," "mary jane," or "weed" are found.	marijuana OR "mary jane" OR "weed"
Co-occurrence (interdependence)	Co-occurring terms must occur along with one (or more) of the primary search terms. In the example, the word "smoke" is a co-occurring term. The page should be scraped and added to the text corpus only if "mary jane" and smoke are found together in the text; if "mary jane" is found alone, then the text should not be added to the corpus. However, the page should be scraped if "marijuana" is found with or without the term "smoke."	("marijuana") OR ("mary jane" + "smoke")
Stop words (or negation or exclusion)	The order of terms that co-occur might also be impacted by word order. If the text contains a stop word, then even if other criteria are met, the text should not be scraped and added to the corpus. In the example, the page should be scraped and added to the corpus if the term "weed" is found in the page, unless the term "dandelion" is also found. If "dandelion" is found, then the page should not be scraped and added to the corpus.	(weed + NOT dandelion)
Proximity	Proximity occurs when two terms are within a certain distance from each other, with the distance usually measured in words. Proximity may apply to both co-occurrence and stop words. The maximum distance may be exact (five words, for example) or using various statistical calculations that take other factors into consideration (length of page, for example). In this example, the word "mary jane" should be within five words of smoke, either in front of "mary jane" in the text or after "mary jane." Five words as the proximity was determined by earlier analysis showing that if the two terms "smoke" and "mary jane" both co-occurred yet were separated by greater than five words, the chance of introducing a false positive into the corpus was higher than what was deemed acceptable error.	example of proximity with co-occurrence: ("mary jane" + "smoke" [proximity: 5])
Negation	All of the components detailed here might be impacted by negation. For example, pull all pages where the word "mary jane" appears unless the term "shoe" also appears (negated co-occurrence).	("mary jane" + NOT "shoe" [proximity: 5])

terms to the search criteria. Certain limitations in scraping the Internet may affect what can be scraped; if a commercial scraping tool is used, the logic for sophisticated term building may not be implemented into the tool. The best approach may be to pull an overly broad textual dataset and remove the false positives within a data preprocessing step prior to analysis. Scraping, or pulling textual data from a webpage, was originally developed for content manipulation, search engine indexing, and other commercial settings. Because of this original purpose, the scraped data may require further refinement that is specific to text analysis.

Given the amount of time that building a search term definition requires for a specific domain, especially when scraping social media data, it is a common mistake to use the filters that were used to scrape or pull the data from the Internet as a classification structure for reporting research results. However, such a structure limits the insight in at least two ways:

- Sentiment analysis accuracy is most likely suboptimal when only search term categorization is used because of a higher likelihood of false-positives within the dataset.
- Clustering techniques are not used on filtering, so unknown categories are not understood.

Deciding which sites to scrape is also important. As an example, if the anticipated research is for a topic with specific relevance among U.S. Hispanics, scraping comments from El Universal.com.mx, ranked 12th in Mexico and with a large U.S. following, should be part of the scraping plan. If not, the textual data may not fully represent the topic domain or come close to approximating the population of those who are discussing the topic.

The time period of the text corpus is also an essential part of scraping design criteria. Some topic domains are seasonal, and for the text corpus to be complete, a full year might be required to cover the topic domain fully enough for analysis. Other topic domains may not require such a long duration. Some social media sites only provide access to historical data (not real-time data) for a short duration. For example, Twitter only allows for a small number of weeks in terms of historical data scrape; after that period, Twitter does not guarantee that data are fully available. Some companies "bank" history of Tweets, and these sites are invaluable when an issue is analyzed over time. From a

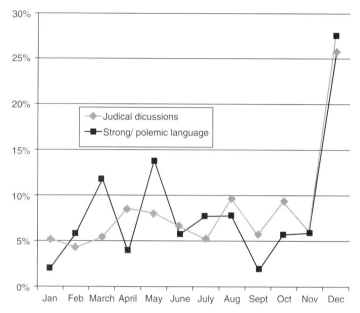

FIGURE 2.4   Spikes on Twitter about health-care reform.

sample of 40,000 Twitter posts over a 25-month period purchased from Radian 6, these posts about the Patient Protection and Affordable Care Act (ACA) were categorized into two categories: "Judicial Discussions" and "Strong/Polemic Language." The first spike occurred in March when the ACA was passed; in December, discussions about judicial rulings spiked when a federal judge ruled that a central component of the ACA violated the Constitution. Figure 2.4 shows a graphic illustration of these spikes in Tweets.

Not all sites consistently return the same data, especially historical or real-time data, even if the scrape is repeated within minutes. One approach to ameliorate this limitation of search and retrieval is to create consistent use of terms and sites scraped rather than trying to achieve a comprehensive scrape of data.

The textual data returned from a scrape possibly have additional metadata information about the site and page from which text was scraped, the date that the page was updated or published, keywords about the text that are valuable to the site, the person who did the posting, the URL where the posting occurred, and other artifacts about the post itself. If a commercial scraping service is used for the scraping, attributes about the poster, such as gender, age, or region of the world, may be imputed based on specific phrases and site information.

Different features of the text corpus before analysis determine the necessary preprocessing of the data to prepare for sentiment analysis. Social media data may need different preprocessing than open-ended survey verbatims. Accompanying images and video may require rescraping for comments and other ancillary information; social media data may have the input mode (mobile, desktop) available as a structured attribute. The specific footprint of the data, such as the way language is used within the text data, informs the sentiment analysis algorithm and models. Specific characteristics to examine are the length of the post; the formality of the authors in writing the post; or whether multiple languages were blended together in the text. In addition, audio that has been translated into text (Twitter posts made by iPhone Siri, for example) may also require special handling. Besides accuracy problems in voice-to-text transcriptions, people use different phrasing in speaking than in writing.

Structure and Understand

NLP is a machine-based (or semi-machine-based) way to prepare textual data so that machine-based algorithms may be applied to the data, enabling computers to make sense of digital text. The algorithms inform the NLP techniques used to preprocess the textual data, as well as use NLP techniques when doing the actual analysis. One preprocessing approach is the ability to "stem" a word, i.e., finding a word's root form, such as removing plurality or tense from a word. Many algorithms exist for stemming; some algorithms apply rules to the word to find its stem, whereas others use lookup tables to find the root form. Another complementary approach is the ability to create n-grams and associate those n-grams with existing text. An n-gram is a sequence of words where n is the number of words that *should* be found in sequence. For example, in the sentence "I am *not* furious," if the data are processed into 2-grams, then the grams would be "I am," "am not," "not furious," as well as the 1-grams, "I," "am," "not," "furious," thus, creating a term list of seven terms altogether. These "new" 2-gram terms, or phrases, may be used in the actual algorithmic analysis. This technique is often called a "bag of words," where the words have associations and types of frequencies attached to the specific words "in the bag."

Other common NLP preprocessing techniques include reducing upper/lowercase words to a single case structure and running statistical frequencies on the sentence, high-confidence associations, and others. Depending on the type of text data under analysis, some of these preprocessing techniques may not be used. For example, with short social media data (like Twitter), depending on the subject, intrinsic meaning may be found in the capitalization of the text—"ma" could mean "mother," or "MA" could mean "Massachusetts," depending on the context of the word in the rest of the sentence. To apply many of these techniques, the text under analysis has to be tokenized, or separated in some way so that it may be parsed for meaning. Usually, spaces between words are used as tokens. In some cases, punctuation marks are used as tokens instead of spaces—thus creating organic grams and not considering individual words as specific items for analysis. Another NLP preprocessing approach is to reduce the textual data so that only those words that give value to sentiment determination are left. This approach reduces the complexity of the algorithms that are run.

APPROACHES TO DETERMINING SENTIMENT

Machine-Coded Sentiment Analysis

Once each post or verbatim within a textual dataset is preprocessed to a specific standard, posts can be compared with one another and analyzed to find similarities and differences. The analysis algorithms used to do the comparison in textual processing determine the methods used for preprocessing text described previously.

The choice of the algorithm to use depends on several factors. How similar are the posts to one another? Do the terms used in the posts vary widely (use many different words), or do the posts use the same terms with high frequency? Do most of the posts have a specialized purpose—for example, are the posts all about movie reviews or product reviews of vehicles? Or are the posts all about a specific topic—alcohol drinking or swimsuit buying, for example? In a study mining subjective data on the web, authors Mikalai Tsytsarau and Themis Palpanas (2012) from the University of Trento in Italy identified over 40 different algorithms developed to determine sentiment from academic literature. These algorithms took the type of textual data into consideration as well as the scalability of the algorithm to work on different-sized datasets.

The use of online data introduces some challenges in both acquiring and managing the textual analysis expectations imposed on the corpus itself. The overall topic domain determined by scraping one or more websites in entirety—without search term filtering—has a direct and possibly great impact on the accuracy of any approach to determine sentiment. If the topic domain of the corpus is wide ranging, for example, covering the subtopics from an entire site such as Facebook, then the resulting sentiment analysis process must be prepared, including significantly more validation.

Some sentiment analysis algorithms use domain- and text-source-specific artifacts, such as custom dictionaries, to increase accuracy, by "tuning" the inputs used to do the polarity determination. These dictionaries are often topic-based. For example, "fine" may be used in a negative sense when discussing hair gel ("the product flaked and left me with fine, limp hair"), whereas the word "fine" may have a positive connotation when a vehicle is discussed ("he looked mighty fine driving that BMW"). The approaches, such as extraction of entities (proper nouns), key noun phrases, and subject classification, have improved the parsing and ultimate classification of textual data.

Note that these examples of algorithms used for sentiment analysis are not exhaustive; other approaches to determining sentiment may be used in both commercial and academic applications. Although the approach to determine sentiment analysis may be algorithmic and repeatable, the algorithm might not necessarily be deterministic, that is, return the same result using the same dataset on multiple executions. Certain algorithms (using nondeterministic algorithms) may exhibit different results under different executions and parameters using the same text.

Methods to determine sentiment analysis break down into supervised learning, semisupervised learning, or unsupervised learning approaches. In supervised learning, a large portion of the posts has already been coded (also called labeled) with a manual sentiment designation. The algorithm or method reads the labeled data and develops a model (using a machine-based algorithm) to use for unlabeled posts. A subset of the total textual corpus is labeled—the rest is used to predict sentiment. (Often an extra step exists—the ability to calculate the precision, recall, and accuracy of the prediction from the model.)

Semisupervised learning (an umbrella term that encompasses classification, clustering, and regression) uses both labeled data and unlabeled data as part of the training. In somewhat simplified terms, using a small amount of labeled data, the process is the same as for supervised learning, except that the labeled data do not have to be as quantitatively large. Instead, the labeled data help jump-start the learning. For sentiment analysis determination, an algorithm is performed independent of the labeled data, and the labeled data then, as they become part of the independent clusters, inform the learning from the cluster—if the labeled data are grouped within the clusters appropriately. Current literature (Zhu, 2008) indicates that semisupervised methods may be as accurate as a fully supervised method with fewer labeled data to use as input. The technique used depends on the type of textual data under analysis: Full documents of text cluster and classify in a different way than social media data, such as Twitter or Facebook posts (or text messages).

Human-Coded Sentiment Analysis

Sometimes, depending on the textual dataset, the simplest approach is the most accurate. Often, the best method for coding a small corpus with no training data is to hand-code the entire textual dataset. This dataset then becomes input for any future dataset with similar properties and topic domains. Note that even humans do not agree when doing manual sentiment analysis coding.

A good training set (manually created labeled data) for sentiment analysis is vital. Labels are a sentiment category assigned to each post in the training set. These labeled data then are used to train a machine to apply sentiment to other posts, or they are used as a validation point for human scorers.

For machine learning, if the labels are incorrect, then the training will be suboptimal, and the scoring will contain errors resulting in poor predictive accuracy of the sentiment. The point of labeled data is to provide input into a supervised or semisupervised method so those algorithms can learn from the posts that have already been assigned a sentiment score. Only a subset of the corpus is hand-coded; the training dataset for semisupervised methods is smaller than what is required from a supervised method.

Labeled data can come from a signal—like an emoticon a sentiment category—or they can come from manually coding. Some researchers use a readied labor force like Amazon Mechanical Turk, or Crowd-Flower, for manual scoring of sentiment. eBay used CrowdFlower to review and assign human sentiment to eBay comments. These tools provide a marketplace of workers who register to receive piecework, called "human intelligence tasks." The rules to do the sentiment coding, for example, the scale used and the criteria in which a specific post would be defined as "negative" or "positive," are written as input criteria for the workers to read prior to coding. Next, a short test is given to verify that the rules were read prior to coding. Finally, the posts that make up the training set are hand-coded independently by two workers. The training data should be cross-validated, that is, hand-coded by more than one worker; then the two sets of codes are compared to determine where they differ. This exercise improves the accuracy of the training data and is paramount to the success of the analysis. If the training dataset (the manually coded posts) is used as input to an algorithm that will code the rest of the dataset, then the final product most likely will have the same quality issues as the training dataset. Thus, independent validation is critical, as reviewing an already assigned score may impact the bias of the person who is doing the coding.

Labeled data may also already exist—if reviews, for example, are available using the same topic domain that is under analysis, and if the reviews may already have a rating system already provided (such as a rating scale of five stars), then those ratings may provide the label necessary for the training of the data.

SENTIMENT ANALYSIS AS A SUBSET OF TEXT ANALYTICS

Sentiment analysis is a part of a larger area of application called text analytics or text mining; both are roughly synonymous (although text analytics is more commonly used today). Text analytics has two techniques worth noting because they are often used with sentiment analysis and can help discover new information in the data that can enhance sentiment analysis. The two techniques are classification into known categories and clustering into new categories based on the organic topics, possibly unknown, within the corpus. To maximize accuracy, both techniques require data preparation and an iterative approach when working with the models.

For purposes of using classification and clustering on social media data, short individual posts are both good and bad for accuracy. Short posts are good because the number of topics in a single post is limited (because of the length); conversely, they are bad because much of the topic context is not available to inform the algorithm.

Classification is finding the similarity between specific posts by comparing those posts with a preexisting (known) categorization scheme and then coding those posts into those categories. A category scheme may have the following appearance:

Swimsuits
  Bikinis
  Cover-ups
  Tankinis
Sweaters
  Cardigans
  Pullovers
  Vests

Classification often involves initially setting up manual rules that define how a single post is coded into a category, using a training dataset of posts. The manual rules use term definition (negation, co-occurrence, stop words, and other techniques) to separate the posts into different categories. Posts may fall into more than one category (a single post may discuss a bikini and a cardigan). Prelabeled data may also be used in place of defining manual rules on a set of training data. The manual rules are in essence creating a labeled dataset without having to hand-code a label for each post. The more sophistication that can be added to the manual rule definitions, such as looking for posts that match a category using term frequency and co-occurrence, the greater the volume of labeled data, and consequently the more accurate the full categorization will be.

After the rules are built (or labels are found), multiple algorithms exist to (at least) semiautomatically classify individual posts into a category (depending on the algorithm used, more than one post may classify into more than one category). After the rules are defined, a good process uses NLP techniques such as automated stemming, synonym identification, and the like for processing the posts to improve accuracy. In a supervised approach, first training occurs, using the labeled data created with the manual rules; then it tests on more labeled data, giving performance results (comparing predicted category with actual labeled data). Finally, the created model runs on the rest of the corpus, separating the unlabeled posts into different categories and completing the classification.

Classification provides the capability to run volumetrics (number of posts that fall in a specific category) on the categorized data, looking at changes over time. The classification scheme may be flat (for example, two categories coded, swimsuits and sweaters) or hierarchical (swimsuits that subcategorize into bikinis, tankinis, and cover-ups).

Clustering is the capability to extract themes from the data itself, without using a predetermined categorization tree. In this instance, themes are determined by clustering the posts by textual attributes in the posts themselves. Labeled data are not used. One approach is first to calculate a similarity measure between each post to see which posts are more like each other than others (most memes share a core set of words that are identical with some dissimilarities between the posts overall). This comparison may help to remove textual memes from the data so that clustering focuses on interesting themes. Another preprocessing step is to remove "function words" or words so common to the data that they do not add value to the cluster itself. The K-Means algorithm is often used to determine the clusters themselves. This algorithm uses the identification of the relative frequency of the terms used in the posts as ways to cluster the different posts in like groups. These like groups may be identified by the words used most frequently within that group.

Clustering provides the capability to identify unknown themes in the data that sentiment analysis and classification by themselves cannot identify. Interestingly, at a technical level, sentiment analysis is a form of text classification, with sentiment categories (positive, negative) as the classification instead of topic-based categories. In combination with other text analytic techniques, sentiment analysis is a powerful filter and organizer and helps highlight information previously unknown. For example, looking at posts within the swimsuit category with negative sentiment as a filter shows that the sizing theme is a new cluster, especially for those posts from females. A subsequent approach is to validate the possible issue with quantitative survey research.

CURRENT LIMITATIONS OF SENTIMENT ANALYSIS

Human beings, when given a corpus of textual data, do not agree completely on the classification of the verbatims or posts (Chew & Eysenbach, 2010). In fact, literature about interratings of classification ranges widely; depending on the content, some multiple reviewers agree 70–79% of the time (Ogneva, 2010). With well-understood and prepared data, a tight topic domain, identifiable and reliable signals, tuned models that encompass a number of approaches, and ongoing validation, semiautomated approaches to sentiment scoring currently can reach 70–79% accuracy (accuracy as validated by human beings; Ogneva, 2010). A few exceptions to these levels of accuracy should be noted. For example, when the author scores his or her sentiment when entering in a textual comment, that score becomes highly predictive to aid in sentiment analysis. If, to extend the example further, the author uses suggested keywords or phrases to aid in entering textual comments, accuracy is also improved, with the keywords mapping to a specific sentiment.

These edge cases aside, unresolved problems persist in keeping levels of accuracy greater than 75% from being achieved, no matter the quality of the analysis. Beware of sentiment analysis snake-oil reports of accuracy levels higher than 80%. Without a tightly controlled topic domain, the size and breadth of social media data and common use of new words and abbreviations affects the level of accuracy in both sentiment analysis and classification.

Author intent and reader perception are often fraught with miscommunication and error. For example, the reader may mistakenly perceive sentiment about a brand when none is expressed. A recent Twitter post from a U.S. poster reads as follows: *"[I] ordered a coke and she brought s [sic] bottle so I asked for a cup and straw lol."* We may think that the writer may like the Coca-Cola brand and is expressing a positive opinion about Coca-Cola (as opposed to another soft drink) because the writer chose to purchase that brand of soft drink. However, "Coke" is not synonymous with the brand "Coca-Cola" in all parts of the United States. "Coke" is often used in the South or the West Coast to mean a carbonated beverage in general, not necessarily the Coke product within the Coca-Cola brand. The writer may have actually ordered a Pepsi, RC, or Red Bull carbonated drink and did not express an opinion about any particular brand but rather carbonated beverages in general. Another potential meaning might allude to the drug cocaine, given the writer's mention of the word "straw" followed by "lol" (laugh out loud). As an aside, these examples also show the limitation of relying on these types of data for traditional social research analysis. An interviewer-administered survey allows for specific framing of a question or for a follow-up probe of a response.

Sarcasm proves to be beyond the grasp of most sentiment analysis approaches unless the problem domain is highly controlled. Anecdotal evidence shows that social media textual data have increased levels of sarcasm and humor used to express sentiment. Examples include "great for insomniacs" when reviewing a book or "where am I?" when commenting on a GPS device. Most sarcasm cannot be detected by any existing system in commercial settings. An exception is research by Tsur, Davidov, and Rappoport (2010), which showed that sarcasm, along with a large amount of labeled data, in a tightly controlled topic domain (reviews) used in a semisupervised algorithm, can be detected with over 70% accuracy.

Inflated expectations of what automated sentiment analysis can produce also impact quality. If sentiment analysis is not validated by external review of the results or manual labeling, the results will also not improve—the reported numbers should not be taken without validation. A sentiment analysis is not accurate because the vendor says it is accurate. Instead, accuracy is measured by validation of the third-party analysis of a significant portion of the coded corpus. A quick validation of a small number of posts is not sufficient to evaluate how well the scoring performed. The same analysis techniques across subject domains do not improve accuracy; each new corpus of data must be reviewed for its unique features that maximize accuracy.

Despite these limitations, sentiment analysis is a valuable tool in helping to find important and unknown research findings buried in large, uncoded text data. Sentiment analysis of social media data has its own constraints and limitations; however, insights validated by survey research allow for significant new knowledge. As long as expectations are properly set and both time and energy are allocated, sentiment analysis may be used to discover insights into textual data that would have gone unnoticed because of volume and complexity.

REFERENCES

CHEW, C., & EYSENBACH, G. (2010). Pandemics in the age of Twitter: Content analysis of Tweets during the 2009 H1N1 outbreak. *PLoS ONE*, *5*(11), e14118.

MOBERLY, T. (2011). *Text-me-up*. London, UK: Beautiful Books.

MOILANEN, K., & PULMAN, S. (2007). *Sentiment composition*. Cambridge, UK: Oxford University Computing Laboratory.

OGNEVA, M. (2010, April 19). *How companies can use sentiment analysis to improve their business*. Retrieved from http://mashable.com/2010/04/19/sentiment-analysis/.

PETERSON, L. (2010, September 7). *Montana author writes entire novel on Facebook*. Retrieved from http://www.prweb.com/releases/2010/09/prweb4466674.htm.

TSUR, O., DAVIDOV, D., & RAPPOPORT, A. (2010). *A great catchy name: Semi-supervised recognition of sarcastic sentences in product reviews*. Presented at the Fourth International AAAI Conference on Weblogs and Social Media (ICWSM 2010).

TSYTSARAU, M., & PALPANAS, T. (2012). Survey on mining subjective data on the web. *Data Mining & Knowledge Discovery*, *24*, 478–514.

ZHU, X. (2008). *Semisupervised learning literature survey*. Madison: Computer Sciences, University of Wisconsin—Madison.

CHAPTER 3

Can Tweets Replace Polls?
A U.S. Health-Care Reform
Case Study

**Annice Kim, Joe Murphy, Ashley Richards,
and Heather Hansen,** *RTI International*

Rebecca Powell, *University of Nebraska*

Carol Haney, *Toluna*

Social media are a powerful source of information that can provide insights on what is happening in the world, what people are thinking, and how information is shared across social networks. With social networks like Twitter enabling mass sharing of information and opinions by users worldwide, researchers have a unique opportunity to harvest these data to understand trends in public opinion where survey data have traditionally been employed. Currently, 85% of American adults use the Internet (Pew Internet & American Life, 2012a) and 66% visit social networking sites such as Facebook and Twitter that enable users to set up a profile and interact with a network of other users on the site (Pew Internet & American Life, 2012b). Among teens ages 12–17, 95% are online and 80% have a profile on a social networking site; social networking is the online activity they spend the most time on daily (Kaiser Family

Social Media, Sociality, and Survey Research, First Edition.
Edited by Craig A. Hill, Elizabeth Dean, and Joe Murphy.

Foundation, 2010; Pew Internet & American Life, 2012c). As these trends suggest, social networking sites have grown in popularity in recent years; Twitter grew from 8 million users in 2009 to more than 140 million users in 2012, generating more than 340 million Twitter postings ("Tweets") daily (Twitter, 2012).

With the proliferation of user-generated content on social networking sites like Twitter, researchers are increasingly analyzing this data to understand what issues are important and what people think about them as well as monitoring emerging trends and threats. Recent studies have demonstrated that people's information-sharing behavior on Twitter is correlated with such phenomena as flu outbreaks, movie box office sales, and opinions about politics and the economy (Chew & Eysenbach, 2010; Signorini et al., 2011; Asur & Huberman, 2010; O'Connor et al., 2010). For example, in their analysis of 1 billion Tweets from 2008 to 2009, O'Connor and colleagues (2010) found that sentiment of Tweets about consumer confidence in the economy and approval of President Barack Obama correlated with trends from random digit dialing (RDD) surveys such as Gallup's daily tracking poll. These studies suggest that the emerging fields of infoveillance (automated and continuous analysis of unstructured, free text information available on the Internet) (Eysenbach, 2009) and sentiment analysis (automated computational coding of text to determine whether opinion expressed is positive, neutral, or negative; see Chapter 5, this volume) can produce results that correlate with and even predict those collected through traditional survey methods. Thus, Twitter offers some promise as a publicly available social media data source that researchers can mine to uncover population trends and infer population attitudes.

Analyzing Twitter data is appealing for several reasons. For one, Tweets are relatively inexpensive and quick to analyze compared with traditional surveys. With growing concern related to declining response rates (Curtin et al., 2005; Tortora, 2004) and decreased landline telephone coverage (Blumberg & Luke, 2009), Twitter represents an alternative source of data where people actively share information. The sheer volume of Tweets generated per day provides a glimpse into the thoughts of some subset of the general population. Second, when analyzing social media data, there is no respondent burden because there are no respondents as traditionally defined. Third, researchers may investigate a topic of interest using Twitter data more quickly than a traditional survey approach. Whereas a survey requires sample identification,

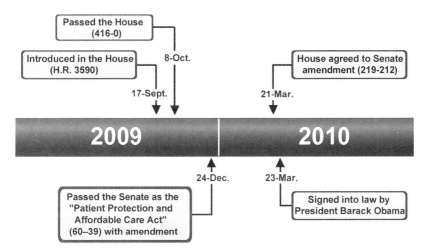

FIGURE 3.1   ACA timeline.

question construction, contact attempts, and data collection prior to analysis, analyzing Twitter data requires only access to the stream of queries and postings, and methods for analyzing the content. Therefore, Twitter may provide an inexpensive option for gaining insight into public perceptions, population behaviors, or emerging topics more quickly than via surveys.

This chapter investigates the utility of Twitter for tracking public opinion about health-care reform. On March 23, 2010, President Obama signed into law the Patient Protection and Affordable Care Act (PPACA or ACA), a comprehensive health-care reform initiative (see Figure 3.1). The bill included extensive provisions to expand coverage, control health-care costs, and improve the health-care delivery system. The provisions had implications for individuals, employers, small businesses, health insurance companies, government health agencies, and public programs such as Medicare and Medicaid. Many components of the law were politically controversial and widely contested in public discourse (e.g., requiring all Americans to purchase health insurance coverage or pay a penalty; expanding Medicaid coverage; and requiring businesses with 50 or more employees to provide health insurance). The Congressional Budget Office estimated that the new health-care reform law would provide coverage to an additional 32 million people when fully implemented in 2019 at a cost of $938 billion, which would be financed through savings from Medicare and Medicaid and new taxes and fees, including an excise tax on high-cost insurance (Kaiser Family Foundation, 2012). For these reasons, traditional polls were conducted during this period to gauge Americans' opinions about the bill. Because of the ACA's wide-ranging effects, information, criticisms, and support for the bill's provisions and changes were likely shared via Twitter in the months leading up to and after its passage. In this chapter, we consider whether Twitter data may supplement traditional survey data for a topic such as public opinion about health-care reform that is of wide interest to the general public, heavily discussed publicly, and likely to shift over time.

The purpose of this study was to examine whether trends in Tweets about health-care reform correlate with observed trends in public opinion about health-care reform. Specifically, we sought to answer:

1. To what extent was health-care reform discussed on Twitter?
2. What is the distribution of sentiment of health-care reform Tweets?
3. Do trends in the sentiment of Tweets about health-care reform correlate with observed trends in public opinion about health-care reform from nationally representative probability-based surveys?
4. What are the key topics discussed in health-care reform Tweets?

We conclude with a discussion on the key findings, strengths and limitations of the study, and areas for future research.

METHODS

Twitter Data

Tweets about health-care reform were identified using Radian6, a leading provider of social media data. Radian6's partnership with Twitter has allowed the company to provide "firehose" access (100% sample) to all public Tweets since May 2008. To identify Tweets about health-care reform, we developed search criteria using the following 20 key terms:

1. health care policy
2. health care reform
3. health care reform bill

4. health care reform law

5. health policy

6. health reform

7. health reform bill

8. health reform law

9. health reform policy

10. healthcare policy

11. healthcare reform

12. healthcare reform bill

13. healthcare reform law

14. obama care

15. obamacare

16. patient protection and affordable care act

17. ppaca

18. us health care reform

19. us health reform

20. us healthcare reform

We searched for health-care reform related Tweets 12 months prior (March 2009–February 2010) and 12 months after (April 2010–March 2011) the signing of the ACA (March 2010). A total of 1,863,607 health-care-reform–related Tweets posted by U.S. users in English were identified for this time period (March 2009–March 2011). We downloaded data from Radian6 that included the actual Tweet (maximum 140 characters), Twitter handle (author), date Tweet was posted, number of followers, number of users followed by the author, and sentiment of Tweet (positive, negative, or neutral) as determined by Radian6.

Sentiment Coding of Tweets

Approach. Sentiment of Tweets is commonly coded using text analysis software that employs defined algorithms to code strings of text automatically. Automated text analysis methods typically require a substantial time investment to set up the coding algorithms, but once they have been established, large volumes of text data can be coded

quickly and consistently. Although these functionalities are advantageous in abstracting high-level information like key topics from volumes
of text, automated text analysis may not be a good fit for analyzing the
sentiment of social media data, especially Twitter, because these data
contain inconsistent and unconventional uses of language (e.g., emoticons, hashtags, and slang) and can have rampant sarcasm. Indeed, in a
previous study, we compared manual coding of Tweets with automated
sentiment analysis methods and found that the accuracy of automated
methods was poor (Kim et al., 2012). We compared the automated sentiment of Tweets provided by Radian6 with manual coding and found
that Radian6's automated coding of sentiment corresponded to manual
coding for only 3% of negative and positive posts.

In contrast, human coders can process tasks like reading Tweets
and picking up sentiment more efficiently than complex algorithms
that may not adequately capture the nuances of human communication.
Manual coding of text data by trained human coders is the traditional
gold standard for analyzing text data, but the massive volume of social
media data can make this process excessively time-consuming and cost-
prohibitive. As a result of these challenges, crowdsourcing these types
of tasks to a wide network of human workers has emerged as a more
cost-efficient solution in analyzing social media data.

Crowdsourcing is the act of outsourcing tasks to a large network
of people who are not geographically bound, which reduces the time to
complete discrete but large-scale tasks (Dalal et al., 2011). Crowdsourcing leverages human capital by finding workers within the network who
can complete the tasks via web/e-mail on their own time and get paid
only for the work completed. We compared multiple crowdsourcing
service companies and chose CrowdFlower because of their experience
crowdsourcing Twitter sentiment analysis and their quality assurance
system, which includes:

- Testing coders' accuracy by interspersing test samples of already
  coded Tweets throughout the coding process
- Using multiple coders per Tweet to help reduce individual bias
- Restricting participation to only those coders with high accuracy
  and deleting and reallocating Tweets of low-accuracy coders to
  other workers

Sample. Instead of sending all 2 million Tweets to CrowdFlower to be coded, we chose a random sample of 1,500 Tweets per month from March 2009 through March 2011. We chose this sample size based on the availability of resources for the crowdsourced coding and because it was in the range of sample size for our comparison with the monthly tracking survey. To draw this sample, we first took into account the users who Tweeted more than once per month. We used only one randomly selected Tweet per unique user per month to prevent the opinions of highly active Tweeters on this single topic from being overrepresented in the sample of Tweets. After we compiled that dataset of one Tweet per user per month, we then drew a random sample of 1,500 Tweets per month. We then sent all 37,500 Tweets (1,500 Tweets for 25 months) to CrowdFlower to code the sentiment through crowdsourcing.

Measures. Before beginning the coding process, the coders were given brief background information on health-care reform and a code-book explaining how Tweets should be coded for negative, positive, or neutral sentiment and why. For each Tweet, coders were then asked to determine whether: (1) it was written in English; (2) it was relevant to the topic of health-care reform; and (3) if relevant, to code the Tweet author's sentiment on a 5-point Likert scale consisting of very negative = 1, slightly negative = 2, neutral = 3, slightly positive = 4, and very positive = 5.

Coding Sentiment. To improve the accuracy of coding, three to six CrowdFlower coders coded each Tweet. CrowdFlower's internal research has found that using three to six coders achieves the optimal accuracy versus cost tradeoff (Oleson et al., 2011). When the coders disagreed, a Tweet was coded by more than three coders. Even when the coders agreed, some Tweets were coded by more than the minimum of three coders because of CrowdFlower's work allocation system.

We also provided approximately 100 Tweets that were not part of the 37,500 study sample as a test sample for coders. These Tweets were coded manually by two trained research team staff (gold standard) and were to be used to assess the reliability of CrowdFlower's coders. The gold standard Tweets were interspersed throughout the sample to monitor the coders continuously. All coders were tested with these gold standard Tweets and given a "trust score" based on their performance.

Only coders with high trust scores coded the study sample of Tweets. Those who performed poorly were no longer allowed to work on the task, and their assigned codes were dropped and recoded by another coder. A total of 1,560 coders contributed to coding the 37,500 Tweets; 9,225 coders were not allowed into the task or were subsequently dropped from it because they did not pass the gold standard test. The overall level of agreement between CrowdFlower's crowdsourced coding and RTI-trained manual coding was high (89.7%) and substantially better than agreement between Radian6's automated sentiment coding and manual coding, which was only 33%.

Topical Coding of Tweets. To understand the key topics discussed in the health-care reform Tweets, we conducted a clustering and classification analysis using RapidMiner, software that uses data mining and machine learning procedures to identify patterns and meaning from large volumes of data. We used this semiautomated approach because it allowed us to control what gets coded and what meaning is given to Tweets that seem to cluster together, and to optimize the algorithms based on initial runs that would then be processed on the remaining data. This approach was favorable to manual coding of Tweets (which would have been time-intensive given the large volume of data), and to crowdsourcing (which would have required that we review the Tweets in advance to provide workers with a list of topics to code).

We used the 37,500 Tweets that were sampled for sentiment analysis as the sample from which to build the topical categories. The first step was to preprocess ("clean") the data to account for factors such as known acronyms, URLs, and spaces between words. We performed the following steps: (1) deleted extraneous characters or words such as "and" and "the"; (2) retrieved actual URLs if shortened URLs (e.g., bit.ly) were used in the Tweet; (3) processed keywords from URLs; (4) stemmed nouns, adjectives, and adverbs using WordNet Stemmer (e.g., supporting → support); (5) developed and used a dictionary to match related keywords (e.g., "Dem" → "Democrat"; senators' names by state); and (6) transformed text to lowercase.

After preprocessing, we conducted cluster analysis, a computational-based process that looks for commonalities among the Tweets. Multiple properties of Tweets were examined for clustering and categorization, including the text of the Tweet itself, the author, whether the Tweet contained a URL or hashtag, and whether the Tweet was re-Tweeted. These properties were included in the clustering algorithms to identify a potential group of Tweets that may be about the same topic. A k-means clustering algorithm was applied to the input, portioning the different Tweets into 15 clusters as a single run. We experimented with other values for k, but 15 returned at least a portion of significant clusters with semantic meaning that could be labeled with a category. The clusters were then processed to build association rules based on terms that co-occurred frequently within the cluster. We took this step to help efficiently identify the potential, specific semantic category of the cluster by drawing out what was important and similar within the cluster.

Once clusters of Tweets were identified, we reviewed the Tweets that had very close connectivity to one another, identified the category of the cluster (if one was apparent), labeled enough of that cluster manually to confirm accurate labeling, and then moved on to the next cluster. After the Tweets were classified (labeled), these Tweets were then used to classify the others that were similar to the labeled Tweets. Those Tweets were then reviewed for accuracy by taking a sample of the Tweets for manual inspection, and if the classification was accurate, it was given a high score in the classification. The clustering and classification process was iterative.

The classification had two steps. First, the labeled data were preprocessed, and then the results were cross-validated. The labeled data were split into two samples (using stratified sampling) to train a support vector machine learner and to test the learner. The learner was tested by giving each Tweet a predicted category and then by comparing the category to the manually labeled category. This process ran a number of times, optimizing the learner each time. A performance evaluator then calculated the accuracy, precision, and other analytic results on the test data. The optimized support vector machine model was then applied to the unlabeled data. This process created a fully categorized dataset with assumed accuracy between 74% and 94%.

A Tweet could cluster into multiple categories but was classified into one final category based on the strength of the cluster relationship. For example, if a Tweet fell into the Judicial (score $= 0.8$) and Medicare (score $= 0.6$) categories, it was classified as Judicial based on the higher score that signifies a stronger cluster relationship.

PUBLIC OPINION ABOUT HEALTH-CARE REFORM: KAISER HEALTH TRACKING POLL

The Kaiser Health Tracking Poll is a Kaiser Family Foundation (KFF) survey that assesses the American public's perceptions about a range of health-related issues. This survey is conducted monthly among a nationally representative RDD sample of approximately 1,200 (and in some cases 1,500) adults ages 18 and older, living in the United States, and includes both landline and cell phone respondents. The surveys are conducted in English and Spanish by Braun Research, Inc. under the direction of Princeton Survey Research Associates International. The response rate (the American Association for Public Opinion Research's RR3 formula) is a little over 10% for both the landline sample and the cell phone sample, and the margin of sampling error is ±3 percentage points. The data are weighted to balance the sample on key demographics (e.g., age, gender, education, and race) using the latest U.S. Census data.

KFF asked a range of questions about general health-care reform over time, but since we were interested in opinions about health-care reform before and after the signing of the ACA, only two questions were consistently asked every month during this period. They were as follows: (1) "Do you think you and your family will be better off or worse off under the new health reform law, or don't you think it will make much difference?" and (2) "Do you think the country as a whole will be better off or worse off under the new health reform law, or don't you think it will make much difference?" The response categories for both questions were measured as better off, worse off, won't make a difference, and don't know/refused.

ANALYSIS

For Twitter data, we calculated descriptive frequencies and graphed trends in measures over time. We also adjusted the raw number of health-care-reform–related Tweets and Twitter authors to account for the overall growth of Twitter users and Tweet volume over time. To do this, we gathered information on Twitter users and Tweet volume statistics provided by Twitter over multiple time points and estimated the number of daily and monthly Twitter users and Tweet volume assuming a linear increase over time. We then calculated health-care-reform–related Tweets and Tweet authors as a proportion of the total estimated Tweet volume and Twitter users monthly.

For Twitter sentiment analysis, we calculated the proportion of Tweets that were negative, positive, and neutral each month. As each Tweet was coded by multiple coders, we first averaged the numeric values of the codes to obtain a single sentiment value for each Tweet. We rounded these values to the nearest integer and collapsed the 5-point Likert scale into the following categories: negative (values of 1 to 2.4), neutral (2.5 to 3.4), and positive (3.5 to 5.0). The sentiment analysis was conducted on a final sample of 36,598 Tweets (out of the 37,500 Tweets in the study sample), as 886 were identified by the coders to not be about health-care reform and 16 were not in English.

For KFF data, we calculated proportions and 95% confidence intervals for both measures and graphed trends over time. The data were weighted to balance the sample of key sociodemographics and to adjust for household size and landline- versus cell-phone-only users. When comparing the Twitter with the KFF trends, we only looked at the positive and negative sentiment of Tweets and perceptions that family/country would be better or worse off.

RESULTS

RQ1: To What Extent Was Health-Care Reform Discussed on Twitter?

Overall, the number of Tweets related to health-care reform increased from 3,000 in March 2009 to 122,589 in March 2011, with the largest peak occurring in March 2010 (270,156 Tweets), the month when the ACA was passed (Figure 3.2). The number of Twitter users Tweeting about health-care reform (health-care reform authors) followed a similar pattern.

When the raw volume of health-care reform Tweets and authors was adjusted to account for Twitter growth, we see that health-care reform Tweets and authors made up a small proportion of the Twitter universe (<0.1%) (Figure 3.3). Additionally, the peak in March 2010 made up a slightly smaller proportion of all Tweets (0.005%) than in August 2009 (0.008%) even though the raw number of Tweets was approximately

72

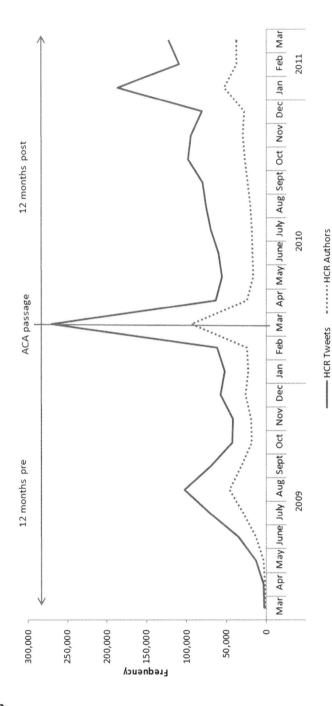

FIGURE 3.2 Number of health-care reform Tweets and number of Twitter users Tweeting about health-care reform (March 2009–March 2011).

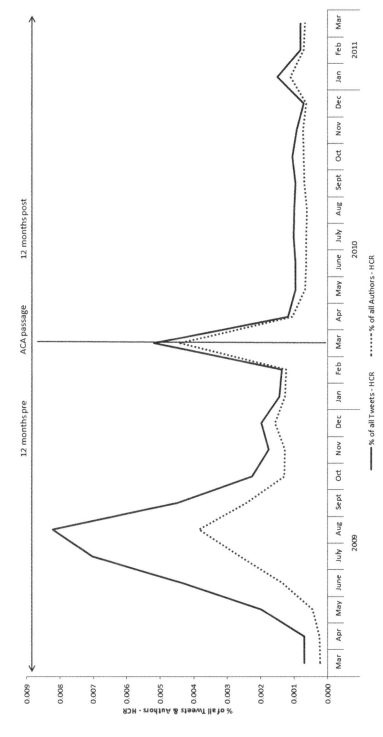

FIGURE 3.3 Adjusted health-care reform Tweets and health-care reform authors as a proportion of all Tweets and Twitter authors (March 2009–March 2011).

2.5 times higher in March 2010, underscoring the growing popularity of Twitter over time.

RQ2: What Is the Distribution of Sentiment of Health-Care Reform Tweets?

As shown in Figure 3.4, most Tweets about health-care reform from March 2009 through March 2011 were neutral. The percentage of neutral (i.e., neither positive nor negative) Tweets decreased slightly over time, from 60% in March 2009 to 45% in March 2011, but the trend line remained relatively stable. The percentage of positive and negative Tweets, on the other hand, changed dramatically over time. In nearly every month, the percentage of Tweets about health-care reform expressing negative sentiment was greater than or approximately equal to the percentage expressing positive sentiment. The gap between percentage positive and negative fluctuated but was relatively minimal in the 12 months pre-ACA passage, but after the ACA passed, the sentiment of health-care reform Tweets was overwhelmingly negative. The greatest

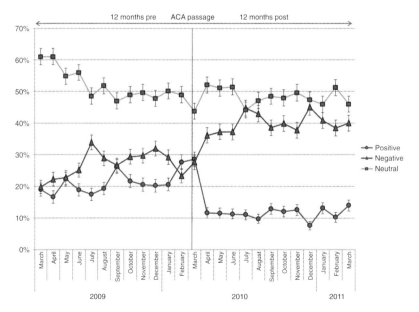

FIGURE 3.4 Positive, negative, or neutral sentiment of Tweets about health-care reform by month (March 2009–March 2011).

difference between positive and negative Tweets was in December 2010 when 45% were negative and only 8% were positive.

RQ3. Do Trends in the Sentiment of Tweets About Health-Care Reform Correlate with Observed Trends in Public Opinion About Health-Care Reform From Nationally Representative Probability-Based Surveys?

KFF TRENDS

In April 2009, 43% of KFF respondents believed the country would be better off under the ACA while only 14% believed the country would be worse off. In every month leading up to passage of the ACA (except January 2010), more respondents said their family would be better off than worse off, but this gap narrowed over time (Figure 3.5).

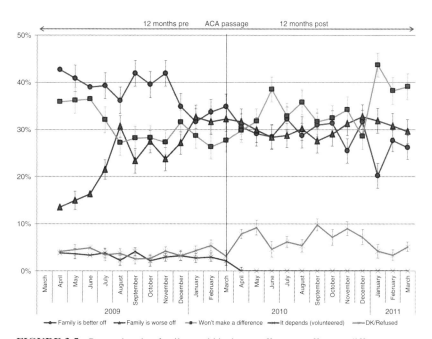

FIGURE 3.5   Perception that family would be better off, worse off, or no different as a result of health-care reform (KFF Survey, March 2009–March 2011). *KFF data were not collected in March and May 2009. There was no statistically significant difference in the 3 months before and after passage of the Patient Protection and Affordable Care Act.

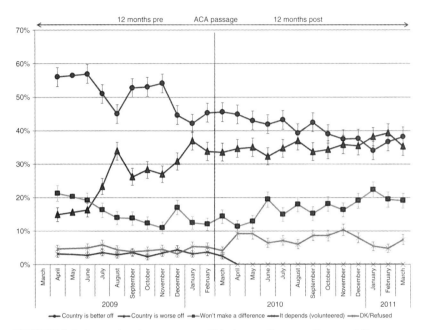

FIGURE 3.6   Perception that country would be better off, worse off, or no different as a result of health-care reform (KFF Survey, April 2009–March 2011). *KFF data were not collected in March and May 2009. There was no statistically significant difference in the 3 months before and after passage of the Patient Protection and Affordable Care Act.

After the ACA passed in March 2010, perceptions that one's family would be better or worse off as a result of health-care reform fluctuated, and beginning in November 2011, more people reported their family would be worse off than better off. Perhaps more telling is that just one month after the ACA passed most respondents thought that the health-care reform would make no difference for their family and this pattern held up for most of the remaining months observed.

Attitudes were much more divided when comparing perceptions that the country, rather than one's family, would be better off or worse off because of health-care reform (Figure 3.6). In April 2009, 56% believed the country would be better off while only 15% believed the country would be worse off. In every month leading up to passage of the ACA, more people believed the country would be better off than worse off, but this gap narrowed over time. Although this pattern held after the ACA passage, the gap narrowed even more over time such that in the first quarter of 2011, no statistically significant difference was evident between these beliefs.

COMPARISON

We compared the KFF and Twitter data to see whether trends in the sentiment of Tweets about health-care reform corresponded with actual trends in public opinion. As Figure 3.7 shows, trends in the sentiment of health-care reform Tweets did not correspond with trends in beliefs that one's family would be better off as a result of health-care reform. On Twitter, the sentiment was consistently negative over time with a more pronounced negative slant after the ACA passage, whereas the KFF data show that the public generally thought that health-care reform would have a positive impact on their family but less so after the ACA passed. Twitter was more negative and not reflective of public perceptions about the benefit that health-care reform may have had on

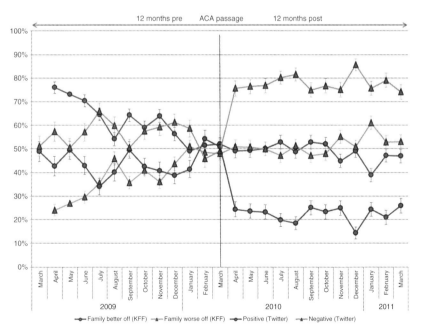

FIGURE 3.7   Comparison of sentiment/support for health-care reform–KFF survey ("family better off") and Twitter (March 2009 - March 2011). *KFF data were not collected in March and May 2009.

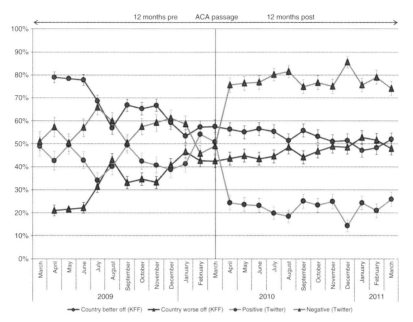

FIGURE 3.8 Comparison of sentiment/support for health-care reform–KFF survey ("country better off") and Twitter (March 2009–March 2011). *KFF data were not collected in March and May 2009.

their families. However, the sentiment expressed on Twitter at times had similar fluctuations as the sentiment in the KFF survey, especially before September 2009, when positive sentiment declined, and after July 2010, when sentiment fluctuated.

The comparison of KFF and Twitter data is similar when beliefs that the country (not one's family) would be better off as a result of health-care reform were considered. As shown in Figure 3.8, sentiment expressed on Twitter was more negative than public perceptions that the country would be better or worse off as a result of health-care reform. However, at times, the fluctuations in the Twitter and KFF data mirrored each other (June–July 2009, August–October 2010, February–March 2011), but the magnitude of these changes was different.

RQ4. What Are the Key Topics Discussed in Health-Care Reform Tweets?

Examining the main topics of Tweets may provide some insights about what issues related to health-care reform may have influenced the

TABLE 3.1

Top 10 Categories of Topics Discussed in Health-Care Reform
Tweets

Topic Categories	Frequency	Percentage
Republican Party/GOP/conservatives mentioned in relation to health-care reform	6,350	16.9%
Specific senator/Congress member/politician mentioned in relation to health-care reform	5,835	15.6%
News/blog/article/poll/website/report/pundit information as a part of health-care reform mention	4,515	12.0%
Judicial ruling or constitutionality under review	4,178	11.1%
Health insurance/Medicare/Medicaid	3,026	8.1%
Congress/House/Senate legislation and passage of health-care reform	2,769	7.4%
Repealing/overturning or defunding health-care reform	1,817	4.8%
Speaking/meeting/event in relation to health-care reform	1,808	4.8%
Paying for/taxes/economy/economics/deficit in relation to health-care reform	1,088	2.9%
Small business/business impact with health-care reform	640	1.7%
All other categories	5,495	14.6%
TOTAL	**37,521**	**100.0 (rounded)**

overwhelming negative sentiment of Tweets. As shown in Table 3.1,
the most common topics mentioned in health-care reform Tweets were
about the Republican Party, GOP, or conservatives (16.9%) followed by
mention of a specific politician by name (15.6%), and then reference to
news articles/blogs about health-care reform (12.0%).

When we examine the sentiment of health-care-reform–related
Tweets for the top topic category, which mentions the Republican
Party/GOP/conservatives (Figure 3.9), we see that these Tweets followed the overall patterns in sentiment observed in the entire sample of
coded Tweets: Most Tweets were neutral, but among those that were
either positive or negative, negative sentiment dominated the entire
period studied, with a dramatic shift in negative sentiment after the
ACA passage.

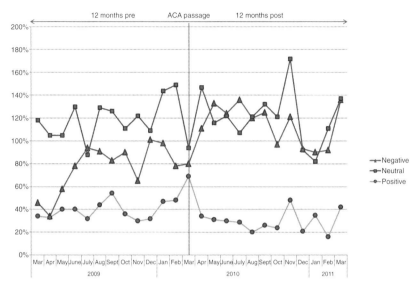

FIGURE 3.9   Sentiment of health-care reform Tweets that mention the Republican Party/GOP/conservatives (March 2009–March 2011).

DISCUSSION

This analysis examined whether the sentiment of conversations about health-care reform on Twitter reflects the trends regarding public support for health-care reform gathered from traditional surveys. We did not have a specific hypothesis regarding the correlation of these two trends; instead, we took an exploratory approach to examine whether these sources may reflect similar levels of positive or negative sentiment about health-care reform in the months before and after the passage of the ACA. Importantly, we were not looking for a "magic correlation" or the best overall fit but whether the sources generally seemed to be providing the same information regarding the trend of opinion about health-care reform.

Overall, we find that the patterns on Twitter do not mirror actual trends in the KFF public opinion poll. Tweets about health-care reform were more negative than positive leading up to the ACA passage, and this gap widened dramatically after the ACA passage. It is likely that the overwhelmingly negative sentiment of Tweets reflected the opinions of political conservatives who opposed the bill; indeed, Republicans/

conservatives and politicians were the leading topics of health-care reform Tweets. An open platform like Twitter enables users to share their opinions about topics that are particularly salient to them. Once the bill passed, supporters likely had no reason to continue Tweeting about the bill, whereas opponents who were angered by the bill's passage may have continued posting negative Tweets about it. In contrast, the public opinion polls reflected more positive support for health-care reform.

Prior to the ACA passage, more Americans believed that the ACA would have a positive impact on their families and the country as a whole, according to the KFF survey. However this optimism declined after the ACA passage; although most people still believed that the country would be better off as a result of health-care reform, more Americans believed that their families would see little/no difference or be worse off. This trend is consistent with previous research from KFF that shows that new proposals are generally very popular when first introduced in Congress, but support usually dies down shortly after people start to learn about what is in the proposal. This trend was observed during the Clinton health-care reform debate in 1993–1994 (Gallup Poll, 1993, 1994). Although we focused primarily on the patterns of positive and negative sentiment, most Tweets were neutral. This finding may suggest that a substantial proportion of Tweets came from individuals and organizations simply sharing the latest news and developments about the health-care reform bill and not expressing an opinion about reform one way or another. Alternatively, we may have overestimated the number of neutral Tweets by only analyzing the Tweet content and not linked URLs or hashtags.

These findings do not paint as promising a picture for simple correlations of survey and social media data as other literature has suggested. For example, O'Connor and colleagues (2010) found that sentiment about consumer confidence and presidential job approvals on Twitter correlate with trends from polling data (Gallup Poll, Consumer Confidence Index, and Index of Consumer Sentiment), in several cases as high as 80%. Similarly, Chew and Eysenbach (2010) found that Tweets about the 2009 H1N1 flu correlated with actual incidence rates of H1N1. In contrast, we found that Tweets about emerging drug topics such as salvia divinorum did not correlate with trends from the National Survey on Drug Use and Health (Murphy et al., 2011). This finding suggests a continuum along which one can expect survey and social media data to produce similar trends.

Trends that are simply a count of a certain type of illness (especially an infectious illness that potentially affects a large proportion of the population) or support for a widely known individual such as the U.S. president may be more straightforward and ripe for comparison, whereas a topic more complex or abstract such as attitudes towards health-care reform may be more difficult for comparing trends across social media and poll data. Factors such as issue salience, social stigma, social desirability, and perceived personal influence may influence the likelihood that someone may Tweet and share a certain opinion on a particular topic. Although little is known about the communication patterns of Twitter users, a topic such as health-care reform may be more difficult to engage in than conversations about flu symptoms or opinions about the president. Furthermore, with only 15% of U.S. adults using Twitter, it is not surprising that Tweets about health-care reform do not correlate with trends from a nationally representative survey like KFF. Twitter's popularity, especially on politically controversial topics like health-care reform, is probably largely driven by organizations, news agencies, politicians, and media/press that use this platform to communicate directly with the public. Although we did not analyze the actual Twitter handles or URLs provided in Tweets to decipher what type of individual or organization was Tweeting, it is likely that news agencies sharing the latest developments in the health-care reform debate and motivated politicians and organizations who were strongly in support of or against the health-care reform vocalized their opinions on Twitter. If true, then Twitter may reflect the opinions of elite media agenda setters rather than those of the general public. As research using Twitter data continues to grow, we may gain better insights on what it is that Twitter trends reflect and in what cases they may supplement or replace traditional surveys.

Our study had several limitations. First, our approach for identifying Tweets about health-care reform may have been biased toward the ACA and it may have missed other Tweets about health-care reform. Because of the nuances of communication on Twitter, identifying the universe of Tweets for any given topic is difficult; with a 140-character limit, abbreviations, hashtags, and slang are pervasive. We did not search on hashtags so we may have systematically missed Tweets in which none of the search keywords were used but the hashtag may have provided clues that the person was Tweeting about health-care reform (e.g., *"So, I got a raise today ... but thanks to the #ACA my insurance benefits went*

up 9.5% #LivingUnderObamaProblems Some people will never learn").
Although we used multiple keywords that were generally about health-care reform and specifically about the ACA, we may have missed other Tweets. In our initial keyword searches we used the more formal description of the bill, Patient Protection and Affordable Care Act (PPACA), and may have missed Tweets that referred to the bill as the Affordable Care Act or ACA.

Second, although we examined multiple measures from the KFF survey that we believed best reflected opinions about health-care reform and would, therefore, be ideal for comparing with Twitter sentiment, respondents' perceptions about the possible impact of the ACA on their families or the country may not reflect their true sentiment or support of health-care reform. We chose these measures because they were consistently collected over several years and we wanted to examine trends over time. We examined two other KFF items that were more closely related to the ACA but asked for a shorter duration ("Given what you know about the health reform law, do you have a favorable or unfavorable opinion of it?" asked since April 2010; "As of right now, do you generally support or generally oppose the health-care proposals being discussed in Congress?" asked from January 2010 to March 2010) and found that these trends did not change our overall conclusions.

Despite these limitations, our study had multiple strengths. First, we examined 12 months before and after ACA passage to look at trends over time. Second, we adjusted the raw Tweet volume and the number of authors to account for the growth of Twitter use over time. Doing so shifted the peaks in the data and underscored the importance of taking into account Twitter growth when conducting trend analysis. Relying on just raw numbers may erroneously suggest that more people were Tweeting about health-care reform or were exposed to related Tweets, when in fact fewer people may have been exposed to health-care-reform–related Tweets because they were competing for attention with an increasingly larger volume of Tweets and Twitter users over time. But we cannot know for certain. For some people who closely follow authors Tweeting about health-care reform or follow specific health-care reform hashtags, the increased volume of Tweets in March 2010 could have dominated their Twitter feed. Having better insights into people's network of followers would allow us to better assess the reach/exposure to health-care reform Tweets.

Third, we tried to control for dominant Twitter authors by randomly selecting only one Tweet from any unique Twitter handle per month. Although this approach gave every person Tweeting about health-care reform an equal opportunity to be selected into our sample, a limitation may be that this approach underestimates the influence of frequent authors who dominated the discourse and therefore may have had a stronger influence on their followers. Future research should compare the benefits and limitations of different sampling approaches.

Fourth, by using crowdsourcing to code sentiment, we obtained more accurate data than relying on automated sentiment analysis from Radian6. When compared with manual coding by trained data collectors (gold standard), crowdsourcing sentiment analysis produced accuracy rates of up to 90% whereas Radian6's automated sentiment analysis accuracy rate was only 3% (Kim et al., 2012). Crowdsourcing may be a more promising method than automated sentiment analysis, but only if the workers can be tested and scored on accuracy and procedures are built in whereby workers with less than a certain level of accuracy can be dropped from the workload and those data points are redistributed to others to be recoded. CrowdFlower employs a multicoder approach and scores coders on accuracy, and as a result, we obtained accurate results on a large sample of Tweets in a timely and cost-efficient manner. Finally, by using Radian6 access to the Twitter firehose (i.e., full data stream), our search results were likely more comprehensive than if we had used other Twitter application program interfaces that only search 1–10% of all Tweets.

CONCLUSIONS

We examined whether data from Twitter and KFF's RDD survey agreed on trends in Americans' view of health-care reform in the months before and after the passage of the ACA bill. Our analysis suggests that Twitter data do not mirror trends gathered from a nationally representative public opinion survey. The sentiment expressed on Twitter toward health-care reform seemed to have been more negative than results from the public opinion poll. Social media data are readily available and relatively inexpensive to gather and analyze compared with traditional RDD surveys. Although they have the potential to supplement survey research, we do not believe, at this point, they can substitute for traditional survey approaches where precise estimates, representativeness, and correlations between standardized measures are needed. Twitter data provide a lens into a segment of the population but exactly who that segment is may vary with topic of study. Future research should examine who uses Twitter and who does not, what motivates people to Tweet, how frequently they do so, and to what extent Tweets are viewed by followers and ultimately what impact, if any, this has on audiences' opinions and behaviors.

REFERENCES

ASUR, S., & HUBERMAN, B. A. (2010). Predicting the future with social media. 2010 IEEE/WIC/ACM International Conference on Intelligence and Intelligent Agent Technology (WI-IAT), 492–499.

BLUMBERG, S. J., & LUKE, J. V. (2009). Reevaluating the need for concern regarding noncoverage bias in landline surveys. *American Journal of Public Health*, *99*(10), 1806–1810.

CHEW, C., & EYSENBACH, G. (2010). Pandemics in the age of Twitter: Content analysis of Tweets during the 2009 H1N1 outbreak. *PLoS ONE*, *5*(11), e14118.

CURTIN, R., PRESSER, S., & SINGER, E. (2005). Changes in telephone survey nonresponse over the past quarter century. *Public Opinion Quarterly*, *69*(1), 87–98.

DALAL, S., KHODYAKOV, D., SRINIVASAN, R., STRAUS, S., & ADAMS, J. (2011). ExpertLens: A system for eliciting opinions from a large pool of non-collocated experts with diverse knowledge. *Technological Forecasting and Social Change*, *78*(8), 1426–1444.

EYSENBACH, G. (2009). Infodemiology and infoveillance: Framework for an emerging set of public health informatics methods to analyze search, communication and publication behavior on the Internet. *Journal of Medical Internet Research*, *11*(1), e11.

GALLUP POLL (1993, September 24–26). [USGALLUP.092993.R03].

GALLUP POLL (1994, April 16–18). USGALLUP.042094.R01].

Kaiser Family Foundation (2010). Generation M2: Media in the lives of 8- to 18-year-olds. Retrieved from http://www.kff.org/entmedia/upload/8010.pdf.

Kaiser Family Foundation (2012). Health reform source. Retrieved from http://sbkbenefits.com/?p=191.

KIM, A., RICHARDS, A. K., MURPHY, J. J., SAGE, A. J., & HANSEN, H. M. (2012). Can automated sentiment analysis of twitter data replace human coding. Presented at the American Association for Public Opinion Research Annual Conference, Orlando, FL.

MURPHY, J. J., KIM, A., HANSEN, H. M., RICHARDS, A. K., AUGUSTINE, C. B., & KROUTIL, L. A. (2011, September). Twitter feeds and Google search query surveillance: Can they supplement survey data collection. Paper presented at the Association for Survey Computing Sixth International Conference, Bristol, IL.

O'Connor, B., Balasubramanyan, R., Routledge, B., & Smith, N. (2010). From Tweets to polls: Linking text sentiment to public opinion time series. Presented at the Fourth International AAAI Conference on Weblogs and Social Media.

Oleson, D., Sorokin, A., Laughlin, G., Hester, V., Le, J., & Biewald, L. (2011). *Programmatic gold: Targeted and scalable quality assurance in crowdsourcing. Human Computation Paper WS-11-11.* Presented at the 2001 Association for the Advancement of Artificial Intelligence Workshop.

Pew Internet & American Life (2012a). Trend data (adults): Demographics of Internet users. Retrieved November 10, 2012, from http://www.pewinternet.org/Trend-Data-(Adults)/Whos-Online.asp.

Pew Internet & American Life (2012b). Trend data (adults): What Internet users do online. Retrieved November 10, 2012, from http://www.pewinternet.org/Trend-Data-(Adults)/Online-Activites-Total.aspx.

Pew Internet & American Life (2012c). Trend data (teens): Demographics of teen Internet users. Retrieved November 10, 2012, from http://www.pewinternet.org/Static-Pages/Trend-Data-(Teens)/Whos-Online.aspx.

Signorini, A., Segre, A. M., & Polgreen, P. M. (2011). The use of Twitter to track levels of disease activity and public concern in the U.S. during the influenza A H1N1 pandemic. *PLoS ONE* 6(5):e19467.

Tortora, R. D. (2004). Response trends in a national random digit dial survey. *Metodolski zvezki, 1*(1), 21–32.

Twitter (2012). What is Twitter? Retrieved from https://business.twitter.com/basics/what-is-twitter/.

CHAPTER 4

The Facebook Platform and the Future of Social Research

Adam Sage, *RTI International*

The fundamental concern of any researcher is data: How do researchers ensure the data tell them what they want to know? Addressing this question in one area often comes at the expense of another. For instance, the survey mode can affect other aspects of a survey such as design, respondent eligibility, and sample frame. Although phone surveys can be more effective than web-based or computer-based surveys in communicating questions and response options, they limit the design and presentation of response options that might enhance the survey experience. For instance, the advantages of a graphical interface (e.g., digital images, slide-bars, and progress bars) on a digital device are difficult or impossible to replicate in a phone survey. Likewise, choosing a web or computer-based survey over a phone survey limits respondents to a certain level of literacy.

Although low or limited literacy does not necessarily affect respondents' ability to listen to questions and respond verbally, it can certainly limit their ability to read instructions, questions, and response options, as well as provide a written response. Digital literacy, or one's ability

Social Media, Sociality, and Survey Research, First Edition.
Edited by Craig A. Hill, Elizabeth Dean, and Joe Murphy.
© 2014 John Wiley & Sons, Inc. Published 2014 by John Wiley & Sons, Inc.

to operate a computer, smartphone, or tablet, can be limiting as well. A web survey intended for a desktop or laptop computer requires a level of digital literacy (e.g., how to operate a mouse, keyboard, or touch screen). In essence, how information is collected always involves tradeoffs. As methods evolve with technology and adapt to recent communication innovations, the costs and benefits of survey mode can come in previously unrecognized and unevaluated forms.

The societal shift to a more digital social existence has left many people wondering about the level and depth at which social phenomena can be understood. Society is moving (and has arguably already made the transition) toward an existence where an individual's identity in a digital environment is more than just an electronic version of real life, but rather it is a new, unique identity created when someone exhibits behaviors solely in a digital environment. One primary digital environment where people exhibit these digitally exclusive behaviors is the social networking site, Facebook. The rapid ascension of Facebook as a website for maintaining social relationships through digital communication, content sharing, and other forms of online interaction has placed the social networking giant in a peculiar position: Facebook faces competing demands to protect its intellectual property and respect the privacy of its users. As the world's largest repository of human data, it is difficult for Facebook to ignore what many feel is an obligation to the research community and society at large to provide access to, or share insights from, that data while ensuring its continued growth and utility to its users and stakeholders. Not surprisingly, Facebook recently garnered the survey research community's attention with questions about its value for evaluating public opinions, attitudes, and behaviors. It is important to understand the magnitude and complexity of Facebook's seemingly ubiquitous existence in society, the information that exists within and about Facebook, and the mechanisms that allow access to this information.

THE CHANGING WEB: FROM SEARCHABLE TO SOCIAL

The arrival of web surveys ushered in new opportunities as well as new restrictions. Methods and best practices were developed to address many limitations (Dillman et al., 1999; Umbach, 2004). As the web became a mainstay in people's lives, the acceptance of web surveys grew along with a concerted effort (1) to understand the web's viability as a survey mode (Fricker et al., 2005), (2) to overcome the challenges associated with a graphical interface for survey administration (Peytchev et al., 2006; Couper et al., 2007; Smyth et al., 2009), and (3) to optimize response rates (Millar & Dillman, 2011). Not only was the web a new mode for administering surveys, but web surveys were also becoming necessary to understand human behavior within the web itself. For instance, asking about customer service satisfaction for online transactions makes more sense as a prompt after a transaction or in a follow-up e-mail and less sense as a phone or mail survey days after a transaction. As the web has evolved, so have web survey methods. Graphical interfaces advanced, download speeds increased, and Internet usage soared, enabling survey researchers to develop more complex surveys with integrated media (photos and videos) that respondents could complete conveniently. Web surveys also benefit researchers because they are more efficient. Recent technological developments have positioned survey researchers at a crossroads between emerging and rapidly developing technologies and traditional, widely accepted, and heavily used survey methods. At this crossroads are social networking sites, especially the current social networking giant Facebook.

In October 2012, more than 1 billion individuals worldwide were monthly active Facebook users. According to Facebook's second quarter filings in June 2012, nearly 57% of these individuals used Facebook on a mobile device (Facebook, 2012a; Facebook 2012b). As of October 2012, in the United States and Canada, 186 million people were Facebook users, and over 65% of U.S. Internet subscribers used Facebook. Figures 4.1 and 4.2 illustrate the gender and age distributions among Facebook users in the United States compared with 2012 Census Population Estimates statistics (Social Breakers, 2012; U.S. Census Bureau, 2013). Generally speaking, females and younger cohorts are more likely to be Facebook users. Although the proportion of female Facebook users is slightly larger than the proportion of females in the 2010 Census, and age distributions do not precisely match Census data, the user statistics follow certain expectations. Specifically, more females than males use Facebook, and young adults are early adopters of new technology. However, although younger cohorts use Facebook in greater proportions, a 2011 Pew survey found the fastest growing age group in terms of social media adoption to be those 65 and older (Madden & Zickhur, 2011).

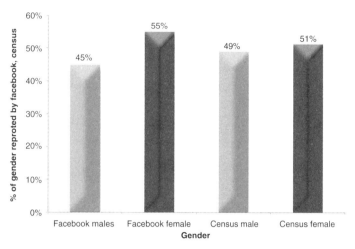

FIGURE 4.1   Gender distribution among Facebook users in the United States compared to 2012 Census Population Estimates.

These 1 billion Facebook users contribute to the 300 million photos uploaded to the site per day, 3.2 billion "likes" and comments per day, and 125 billion friendships, or connections, among people. Users also express opinions and moods via status updates; share a wealth of Internet content (e.g., videos, news stories, blogs, and URLs); join groups; and

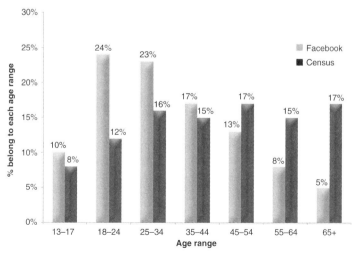

FIGURE 4.2   Age distribution among Facebook users in the United States compared to 2012 Census Population Estimates.

express interest in any entity with a label (e.g., brands, events, music, pop culture, political figures, and ideologies), all while exploring and discovering new content purposefully or via happenstance. The result is that users spend more time on Facebook than on any other website (Nielsen, 2011).

In essence, this communications platform links people to people and people to objects through their behaviors. Facebook calls this network of entities and connections the "social graph." Assuming Facebook's continued dominance of the online social networking market, as Facebook evolves, so will users and their behaviors. The demographic landscape of its users will change and distinctions between current user characteristics (e.g., age) will start to fade, new Facebook products and features will be added (e.g., the searchable social graph), and certain behaviors could become outmoded or change entirely. For instance, if Facebook provides new options to "want" something (a feature tested in 2012), rather than using "like" as the universal expression for approval, more wants of consumer goods may appear near holidays and birthdays, and more likes could appear about sentiments or statuses. Ultimately, these aspects of cultural evolution will define the ways in which platforms such as Facebook are used for research. For instance, supplementing survey data with a list of someone's likes means something entirely different when the want option is introduced.

Each generation is brought up with a preferred mode of communication. For Millennials (colloquially, those born in the 1980s and 1990s and constituting a majority of Facebook users whose values and behaviors are more directly impacted by the technological evolution involving the web), that mode is currently Facebook. Although access to, and volume of, information is unprecedented, much of this information comes in the form of new behaviors. These new behaviors, often in the form of social sharing, have embedded in them much information researchers seek, such as diet and exercise. Whereas shared experience once meant holding a debate-watch party or discussing a newspaper article at the water cooler, it now connotes an entirely new set of behaviors. No longer limited to face-to-face communication, letters, phone calls, or even e-mail exchanges, more people now take part in real-time dialogue during a political debate, for instance, through status updates and comments; liking political figures, ideologies, or a friend's status update about the latest gaffe; sharing voting behavior via photos; or sharing the action of political donations with their Facebook friends.

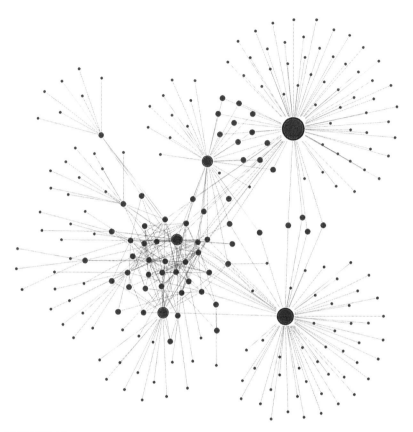

FIGURE 4.3   Facebook's API structure connects datapoints, such as people to people, objects, and events.

When Facebook introduced the Graph API (application programming interface), it provided a gateway for developers to access all of this information within the social graph. Accessing the social graph through the Graph API requires an understanding of the API structure itself. Figure 4.3 shows associations among data points within the social graph vis-à-vis the Graph API structure. Each data point on the social graph (e.g., first name, last name, date of birth, status update, and profile photo) is accessible through a permissions granting entity paired with a data accessing mechanism. Typically, Facebook developers access the social graph via permissions from the user, which then use data to enhance an application experience.

DIGITAL AND DIGITIZED DATA

Using Facebook for any type of data capture first requires consideration of data type. Data-laden experiences native to the platform itself make Facebook unique as a mode of data capture. Before social networking sites like Facebook, standardizing concepts such as "friend" required researchers to provide explicit definitions or to leave an interpretation of the concept for study participants, measuring reliability through statistical analysis. By giving all Facebook "friendships" a ubiquitous characteristic—every Facebook friendship is a social, mutually agreed-upon connection between two individuals—a level of standardization is inherent in its construct. These native features of Facebook make some of the data *digital*. In other words, some phenomena inside Facebook cannot occur or are incredibly difficult to quantify outside Facebook's digital space. Measuring social contacts in the real world to the extent they exist in Facebook would be a tedious task, if not altogether impossible. Likewise, broadcasting a status update or a like in the real world to all one's friends would be downright absurd in most circumstances. Therefore, the connections one has in Facebook allow certain experiences to occur that cannot occur anywhere else. Status updates, comments, shares, and likes are native to the platform: they do not exist outside Facebook in the same way and take on different meaning without the context of Facebook. These communications are *digital data*, occurring and existing exclusively in a digital environment.

On the other side of this data divide are *digitized data*. As we communicate our real-world experiences in Facebook and other digital platforms, we digitize our lives. When I share my age, the last song I listened to, or the distance of my afternoon jog, I am communicating a tangible fact of my real life in a digital environment—I am digitizing my life. Similarly, a survey can be digitized, as with web surveys. Answering questions through a digital interface is a form of digital communication that requires some level of digital literacy. In other words, whether I share the last song I listened to on Facebook by linking my Facebook account with my Spotify account (a music service on the web that connects to Facebook's API to enable sharing of artists and playlists), or I type the last song I listened to in an empty text box in a web survey, I have to understand how to navigate a digital interface to communicate, or digitize, that behavior.

Facebook offers social scientists more than a new mechanism for digitizing behaviors. Although a survey could fit into the Facebook environment in certain instances, much of the potential value of Facebook involves harnessing the experience or the information and processes unique to or uniquely visible through the Facebook platform. This point is important because digitally native experiences are relevant to real life. Scientists at Facebook have analyzed positive and negative sentiments in status updates and have found overall positive sentiments to occur during happier times, such as holidays, and negative sentiments to occur around less happy times, such as natural disasters (Kramer, 2010). A 2012 Pew report also found that 39% of American adults engaged in politics in some form (e.g., liking a post, commenting on a status, or following a politician) on a social networking site (Rainie et al., 2012). So although expression of moods and political engagement both occur outside Facebook, the *way* in which they occur, be it a status update or like, does not occur outside Facebook. As a result, digitally native experiences and the voluntary digitalization of daily life add a dimension to data capture capabilities beyond the traditional survey approach.

Social scientists, including survey researchers, may find value in Facebook users; in other words, individuals are now more than just respondents. They are users and respondents, requiring an approach that goes beyond mere data collection and involves particular considerations to users' experiences throughout their participation. A traditional survey can ascertain through respondent recollection whether a behavior occurred, but it cannot easily ascertain the content or context (when and why) of that engagement (for example, in determining whether an individual was politically engaged on a social networking site). To arrive at an understanding that encompasses both a user characteristic and its content and context, the data collection process must extend beyond a question and answer and tap into a larger platform.

THE CASE FOR FACEBOOK INTEGRATION

Many researchers seeking to integrate Facebook into their work, especially survey research, may intuitively gravitate to Facebook for digitizing data. After all, considering the rise of cell-only households in the United States (Blumberg et al., 2012), and the growth in Facebook users since its inception in 2004, Facebook users could soon become more common than landline users. However, the lack of a sample frame has been a point of concern that has given way to exploratory research in nonprobability sampling methods (Brick, 2012; Bhutta, 2012). The question/dilemma is as follows: How do survey researchers integrate this behemoth of a data source in a way that enhances current methods while adhering to the standards and principles that have always grounded scientific inquiry?

To address this issue, it is important to know what considerations and criticisms apply to research that seeks to incorporate Facebook. As noted earlier, historical approaches to surveys have evolved along with modes of communication. As the telephone survey evolved and the capabilities of touchtone telephones were understood, data collection methods evolved to include self-administered touchtone surveys. Likewise, as data collection devices evolved from paper and pencil, to laptops, and now to tablets and smartphones, the ability to collect certain types of data became an attribute of the device's functionality. For instance, understanding the time and location of an interview evolved from an honor system and ability to perform costly quality control, to an ability to attach specific geographic coordinates and timestamps to questionnaire administration, leaving no doubt as to the exact time and location of the interview. As with the introduction of the touchtone survey, timestamps and geolocation captured by the tablet or smartphone during data collection are capabilities native to the new device, providing a level of improvement from its predecessor. But although researchers certainly strive to improve methods development, sacrifices are often also required. For example, replacing a telephone interviewer with an automated touchtone survey removes a level of assurance that respondents understood and seriously considered questions. And replacing paper-and-pencil surveys with tablets or smartphones introduces new challenges, such as digital literacy, screen size, operating system capabilities, broadband connectivity, and hardware limitations such as battery life.

Every time researchers encounter that crossroad between tradition and innovation, they find that understanding phenomena tied to the functionality of new technology, or understanding how new functionality allows data capture, requires an amalgamation of the old and the new. With Facebook, an amalgamation of digital and digitized data collection is needed, which is now possible through Facebook's Graph API. In sharp contrast to understanding an individual in a survey, Graph

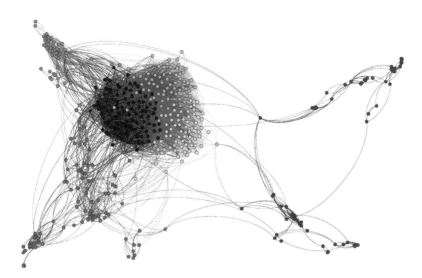

FIGURE 4.4   A personal Facebook network created using Gephi SNA Software (Bastian et al., 2009).

API allows relatively inexpensive and uniquely robust insights into the respondent as a Facebook user.

As Figure 4.4 shows, unraveling the Facebook experience can identify a social context previously unrivaled by any survey. This graph consists of a portion of an individual's personal Facebook network. Each Facebook friend is indicated by a point, or node, and the friendships that exist among friends are indicated by the lines, or ties, connecting each node. Hidden within these networks is a plethora of information waiting to be found via social network analysis. For instance, by running a simple modularity algorithm, whereby communities or clusters of individuals with a high density of shared friendships (compared with a network of random friendships) are identified, context becomes more apparent. The darkest nodes in the center are members of the individual's high-school graduating class. The lighter nodes clustered to the right are friends from other graduating classes from high school. To the bottom left are college friends; further to the left are friends from an internship program. To the top right are co-workers at a current job, and just below are friends from graduate school. (Interestingly, the link between the graduate school cluster and current co-workers is an officemate in graduate school who now happens to work with the user.)

By obtaining such user data about a respondent, one might begin to deduce processes that somehow relate to survey response. Whereas social phenomena are often described by groups of individuals (e.g., lower educated individuals tend to be less politically active than their higher educated counterparts), researchers may begin to recognize new, equally informative patterns (e.g., lesser educated individuals become more politically active as they become socially closer to their more educated counterparts).

DATA AND THE GRAPH API

Whether digital or digitized data are needed, Facebook's Graph API is the gateway between researcher (or developer) and data; Graph API is the mechanism that enables access to the social graph. For years, survey researchers have administered web surveys in a searchable web environment, essentially linking pages to other pages and asking respondents to provide data as they clicked through. As revolutionary as this environment was, the task of obtaining survey data along with network data was still seemingly impossible to accomplish on a large scale. Linking individuals together required a step (or leap) beyond the administration of a standard survey. For instance, the laboriousness and complexity of identifying a respondent's social network, with the exception of perhaps the most local levels (e.g., immediate family members), and the shared relationships within each individual in the network would have been cost-ineffective. Consequently, understanding survey responses in context was an impossible task, and establishing how individuals were connected was still a task akin to the methods of Stanley Milgram's "Small-World Experiment" (1969).

When Milgram set out to explore the social fabric to determine the distance between two random individuals, he developed a series of experiments to measure the average number of ties between two seemingly randomly chosen individuals. In one experiment, letters were sent to individuals in Nebraska instructing each recipient to forward the letter to a friend more likely than themselves to know a randomly chosen individual in Boston. In this way, Milgram could count the average number of hops the letter took to reach a recipient. When a recipient knew the target individual, they were to send it directly to the target, thus, ending the chain and establishing a total number of hops. In what is now commonly referred to as "6 degrees of separation," Milgram found that the average number of hops was 5.2 or just over 6 degrees. Methodological limitations aside (Nebraska and Boston are not necessarily two disparate worlds when associated along a variety of characteristics of social identities, not just geographic distance), the level of effort to arrive at one number is impressive considering that data on Facebook friendships are far more conducive and accessible to measuring social proximity than forwarding letters. In fact, scientists have shown that Facebook is connecting the world, and 92% of all pairs of users are connected within 4 degrees (Backstrom, 2011). In 1969, understanding these phenomena using data from more than 1 billion people was probably inconceivable. Today, scientists can understand social ties on a global scale, and they can now use such analytic capabilities to investigate phenomena more complex than average social distance, such as flu outbreaks (Balcan et al., 2009) and friendship dynamics among college students (Lewis et al., 2008).

Since its creation in a Harvard University dorm room in 2004, Facebook has grown to become a platform that not only connects people to people but also connects and maps people to objects. Throughout its evolution and continuing efforts to connect people, Facebook has introduced several tools for communication, including content sharing, status updates, commenting, liking, blogging, e-mail, instant messaging, and video chat. These efforts, among others, have successfully cemented Facebook as a legitimate supplemental and, for some, outright alternative communication platform. Former Facebook President Sean Parker acknowledged that one clear objective of Facebook early on was to become a new communications platform (Kirkpatrick, 2011). Considering its current size, breadth of utility, and rate of use, Facebook has arguably accomplished this objective. Key trends in communication, such as declining landline use and increasing mobile use and smartphone adoption, have contributed to trends in survey research, such as declining response rates across traditional survey modes and higher response rates with electronic data collection methods (Carley-Baxter et al., 2010; Baruch & Holtom, 2008). Altogether, these trends lend credence to the notion that Facebook is a new, viable, and arguably sustainable communications platform worthy of serious consideration and rigorous scientific investigation.

In exploring the possible uses of Facebook through the Graph API for research, it is important to think of data capture in regard to user engagement. Applications that offer users the ability to connect with friends and engage in a common activity, such as social gaming, create reasons for users to exhibit certain behaviors. In social gaming, the return for the vendor is engaged players, and in some circumstances, engaged players become more engaged when they know their opponent: *Words with Friends* is better when played with friends.

FACEBOOK APPLICATIONS

Accessing the social graph in more depth and developing data capture functions requires programmed processes or applications (i.e., apps). Where the Graph API is the gateway to the social graph, Facebook applications are the keys.

The fundamental difference between a Facebook application and any other web application is access to the social graph. Any website or web application can connect to Facebook via plugins (discussed in following section), but a true Facebook application uses component(s) of the social graph. In this sense, any web survey can easily become a Facebook application, and accessing profile information to prefill certain demographic questions is a simple use of the Graph API. However, simply conducting a traditional survey within Facebook does not offer opportunities much beyond what web surveys currently offer. A true use of applications in survey research is the social survey: understanding the individual in context. Developing such a survey requires an integration of the Facebook experience into the data collection process. Working with this approach requires recognition that some aspects of collection of data will be fundamentally different with Facebook. For instance, incentive structures might evolve into a rewards system or gamified approach (see Chapter 11 in this volume on gamification and survey research). Facebook users interact on Facebook in ways that are not always monetarily driven. Often individuals engage with third-party applications because the reward is a sense of belonging, the task is simply fun or a time-filler, it serves some practical function, or it is part of an acquired identity.

In 2012, I developed an application called Reconnector to test the feasibility of several aspects of a Facebook data collection application (see Figures 4.5, 4.6, and 4.7). Specifically, I aimed to develop a registry of individuals with military affiliations (active duty, veterans, reservists,

Connect with old and new military friends by joining the growing network of active duty, veterans, spouses, brats, and more!   **Connect with Facebook**

FIGURE 4.5   Front page of Reconnector application.

spouses, and children) by providing a service. For the user, a short survey collected background information not available through the Graph API, and the Graph API was used to fill in bits of personal information, such as name, age, friends, interests, and location. Together, these pieces of information were used to create a database that would power advanced searches for application users (at that time Facebook was not capable of doing this search). In 2013, Facebook introduced the Graph Search, which performs search functions of the social graph previously limited to name, location, e-mail, employer, and school. Although the application achieved limited success, I discovered that several aspects of such applications will determine their success.

At the core of any application is the concept that drives its use. Therefore, a useful application enhances some experience (e.g., Facebook). Developing a concept for a Facebook application requires an understanding of Facebook users, not only their demographic composition, but also the behaviors exhibited and the meanings behind their behaviors. For Reconnector, I assumed the military population might

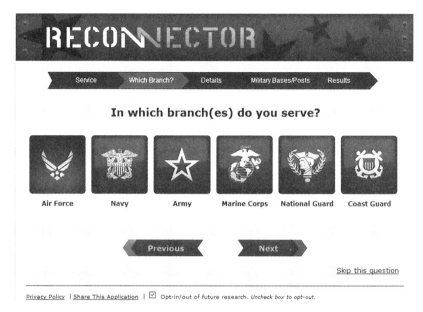

FIGURE 4.6   Example question from Reconnector application.

find an advanced search mechanism useful for a few reasons. First, the military is a very mobile population. Individuals with military affiliations usually relocate from one installation to another. Second, Facebook was introduced in the midst of two major wars, a time likely marked by more frequent mobility among active duty military and their families. Once an application's concept has a firm grasp on these aspects of its potential users, the rest follows—if all goes well.

Depending on what the intended user base looks like, a user-acquisition strategy can take many forms. One unlikely strategy (for acquiring application users) is to have an application go viral (e.g., to have the growth depend solely on the application's ability to appeal to an unknown population). Therefore, knowing the potential user base and the application's potential appeal to it are key. For instance, if the potential user base is from a sample survey, the best strategy may be to recruit users while administering a survey. If the potential user base has certain known characteristics, targeted advertising in Facebook may be best. For Reconnector, the strategy involved a series of targeted paid ads. To develop these ads, I had to understand the population in respect to my application. Because the wars began prior to the introduction of

FIGURE 4.7 Results page of Reconnector application.

Facebook for public use, a rise in the sheer number of enlisted active duty, reservist, and National Guard personnel meant a rise in social connections occurring without the luxury of maintaining them through Facebook. For this reason, I assumed those that were 18 years old prior to public access to Facebook would be more likely to use the application, which was indeed the case. In fact, less than 15% of users were 24 years old or younger.

Finally, monitoring user behavior and developing a responsive and adaptive approach to engaging users ensure an application's vitality. Although some aspects of data collection applications, such as Institutional Review Board protocols and consent processes, are unavoidable, an application should prioritize user engagement. Understanding when and how users engage, and doing so in real time, is a unique opportunity to optimize any data capture effort. For Reconnector, this strategy meant creating a seamless experience, eliminating bugs during the questionnaire process, and presenting intuitive and useful results. For instance, as shown in Figure 4.7, results were provided along with filters that allowed users to narrow results by age range, gender, and key words one might find in other areas of the profile, such as a user's likes and interests.

SOCIAL PLUGINS

Social plugins can also serve as unique methods for engaging respondents and, in some cases, collecting data. Social plugins are designed to create a social web experience, which can expedite building non-probability samples. For example, adding a like button or Facepile (a plugin that shows which Facebook friends are engaged on a third-party website) to a survey allows users to see how many people have engaged with a survey or how many friends have taken the survey. A like button can also be used to generate a snowball sample and lend legitimacy to the survey as respondents will see how many users have liked the survey and will begin to notice friends who liked the survey in their newsfeed as the number of likes increases.

The registration plugin offers the ability to prefill some survey or screener data automatically. Use of the registration plugin not only provides convenience to the user by eliminating question administration time but also provides a level of identity validation. From its inception,

Facebook's design has promoted the use of users' authentic identities. To build real social networks and promote accountability to behaviors, Facebook policy stipulates that each user can have only one account. Although loopholes exist that allow for issues such as multiple accounts for one user, underage accounts, accounts intended for spamming, and "fake" accounts (8.7% of accounts as of June 2012), Facebook has shown the ability to reach up to 95% accuracy in targeting ads (Facebook, 2012a; Facebook, 2012b).

THE FUTURE, MOBILE APPS, AND THE EVER INCREASING COMPLEXITY OF THE SOCIAL GRAPH

Current data on smartphone adoption (Pew, 2012) indicate that most cell-phone owners in the United States own a device capable of carrying mobile apps (Rainie & Fox, 2012). After Facebook's 2012 initial public offering, immediate pressure from investors to monetize the mobile experience was evident to the point of sharp decline in share values. Such a significant response only emphasizes that many believe the social web is becoming more and more of a mobile experience. With this experience comes more opportunities to share and engage, more photos to be uploaded, more check-ins to occur, and more data to be created.

REFERENCES

BACKSTROM, L. (2011) *Anatomy of Facebook*. Retrieved January 24, 2013, from http://www.facebook.com/notes/facebook-data-team/anatomy-of-facebook/10150388519243859.

BALCAN, D., COLIZZA, V., GONÇALVES, B., HU, H., RAMASCO, J. J., & VESPIGNANI, A. (2009). Multiscale mobility networks and the spreading in infectious diseases. *Proceedings for the National Academy of Sciences of the United States of America*, *106*(51), 21484–21489.

BARUCH, Y., & HOLTOM, B. C. (2008). Survey response rate levels and trends in organizational research. *Human Relations*, *61*, 1139–1160.

BASTIAN, M., HEYMANN, S., & JACOMY, M. (2009). *Gephi: An open-source software for exploring and manipulating networks*. Presented at the International AAAI Conference on Weblogs and Social Media.

BHUTTA, C. B. (2012). Not by the book: Facebook as a sampling frame. *Sociological Methods and Research, 41*, 57–88.

BLUMBERG, S. J., LUKE, J. V., GANESH, N., DAVERN, M. E., & BOUDREAUX, M. H. (2012). *Wireless substitution: State-level estimates from the National Health Interview Survey, 2010–2011.* National Center for Health Statistics. Retrieved from http://www.cdc.gov/nchs/data/nhsr/nhsr061.pdf.

BRICK, J. M. (2012). The future of survey sampling. *Public Opinion Quarterly, 75*(5), 861–871.

CARLEY-BAXTER, L. R., PEYTCHEV, A., & BLACK, M. C. (2010). Comparison of cell phone and landline surveys: A design perspective. *Field Methods, 22*, 3–15.

COUPER, M. P., CONRAD, F. G., & TOURANGEAU, R. (2007). Visual context effects in web surveys. *Public Opinion Quarterly, 71*(4), 623–634.

DILLMAN, D. A., TORTORA, R. D., & BOWKER, D. (1999). *Principles of constructing web surveys.* SESRC Technical Report. Pullman, WA: Social and Economic Sciences Research Center, Washington State University.

Facebook (2012a). *Amendment Number 4 to Form S-1.* Retrieved from http://www.sec.gov/Archives/edgar/data/1326801/000119312512175673/d287954ds1a.htm.

Facebook (2012b). *Form 10-Q.* Retrieved from http://www.sec.gov/Archives/edgar/data/1326801/000119312512325997/d371464d10q.htm#tx371464_14.

FRICKER, S., GALESIC, M., TOURANGEAU, R., & YAN, T. (2005). An experimental comparison of web and telephone surveys. *Public Opinion Quarterly, 69*(3), 370–392.

KIRKPATRICK, D. (2011). *The Facebook effect: The inside story of the company that is connecting the world.* New York: Simon & Schuster Paperbacks.

KRAMER, A. D. I. (2010). *How happy are we?* Retrieved from http://blog.facebook.com/blog.php?post=150162112130.

LEWIS, K., KAUFMAN, J., GONZALEZ, M., WIMMER, A., & CHRISTAKIS, N. (2008). Tastes, ties, and time: A new social network dataset using Facebook.com. *Social Networks, 30*, 330–342.

MADDEN, M., & ZICKHUR, K. (2011). *65% of online adults use social networking sites: Women maintain their foothold on SNS use and older Americans are still coming aboard.* Retrieved from http://pewinternet.org/~/media//Files/Reports/2011/PIP-SNS-Update-2011.pdf.

MILGRAM, S. (1969). The small-world problem. *Psychology Today, 1*(1), 61–67.

MILLAR, M., & DILLMAN, D. A. (2011). Improving response to web and mixed-mode surveys. *Public Opinion Quarterly, 75*(2), 249–269.

Nielsen (2011). *Social Media Report: Q3 2011.* Retrieved from http://blog.nielsen.com/nielsenwire/social/.

Pew (2012). *Smartphone update 2012.* Retrieved from http://www.pewinternet.org/Reports/2012/Smartphone-Update-2012/Findings.aspx.

PEYTCHEV, A., COUPER, M. P., MCCABE, S. E., & CRAWFORD, S. D. (2006). Web survey design: Paging versus scrolling. *Public Opinion Quarterly*, *75*(2), 249–269.

RAINIE, L., & FOX, S. (2012). *Just-in-time information through mobile connections.* Retrieved from http://pewinternet.org/~/media//Files/Reports/2012/PIP_Just_In_Time_Info.pdf.

RAINIE, L., SMITH, A., LEHMAN SCHLOZMAN, K., BRADY, H., & VERBA, S. (2012). *Social media and political engagement.* Retrieved from http://www.pewinternet.org/Reports/2012/Political-engagement/Summary-of-Findings.aspx?view=all.

SMYTH, J., DILLMAN, D. A., CHRISTIAN, L. M., & MCBRIDE, M. (2009). Opened-ended questions in web surveys: Can increasing the size of answer boxes and providing extra verbal instructions improve response quality? *Public Opinion Quarterly*, *73*(2), 325–337.

Social Breakers (2012). *United States Facebook statistics.* Retrieved from http://www.socialbakers.com/facebook-statistics/united-states.

UMBACH, P. (2004). Web surveys: best practices. In S. R. Porter (Ed.), *Overcoming survey research problems: New directions for institutional research* (pp. 23–38). San Francisco, CA: Jossey-Bass.

Unites States Census Bureau (2013). Annual Estimates of the Resident Population by Single Year of Age and Sex for the Unites States: April 1, 2010 to July 1, 2012. http://www.census.gov/popest/data/national/asrh/2012/index.html.

CHAPTER 5

Virtual Cognitive Interviewing Using Skype and Second Life

Elizabeth Dean, Brian Head, and Jodi Swicegood,
RTI International

Surveys are prone to numerous sources of error. Survey methodologists refer to the combination of these errors as total survey error (TSE) (Biemer & Lyberg, 2003). Measurement or response error is a major component of TSE and stems from the question-and-answer process (FCSM, 2001; Groves et al., 2004; Groves & Lyberg, 2010). The primary purpose of cognitive interviewing is to identify and minimize measurement error (Willis, 2005; Beatty & Willis, 2007). Cognitive interviews have been used to help identify question comprehension problems (Harris-Kojetin et al., 1999; Beatty, 2002), vague or inconsistent instructions (Carbone, Cambell, & Honess-Moreale, 2002; Zuckerberg & Lee, 1997); problems with logic or flow (Redline et al., 1998; Harris-Kojetin et al., 1999); and issues with participant burden or questionnaire complexity (Beatty, 2002; Miller, 2003).

Over the past 30 years, cognitive interviewing has become one of the most commonly used questionnaire pretesting techniques among academic, government, and private survey research organizations. For

Social Media, Sociality, and Survey Research, First Edition.
Edited by Craig A. Hill, Elizabeth Dean, and Joe Murphy.
© 2014 John Wiley & Sons, Inc. Published 2014 by John Wiley & Sons, Inc.

much of its history, cognitive interviewing has been conducted in laboratories housed at these institutions. Advantages to hosting interviews in a laboratory include control over the interviewing environment, the capability to use technologies that require proximity to the laboratory (e.g., video recording interviews), and scheduling convenience for the interviewing team. Disadvantages include the inability to recruit and interview geographically dispersed populations, inconvenience for participants who must travel to the laboratory, and the artificial nature of a laboratory interview. Recent technological developments allow survey methodologists to conduct cognitive interviews through computer videoconferencing software (e.g., Skype videoconferencing) and virtual worlds (via avatars). This chapter presents a brief background on how cognitive interviewing developed and how it is commonly used today, and it evaluates some new modes that facilitate the recruitment of participants and the interviewing process itself.

BRIEF BACKGROUND ON COGNITIVE INTERVIEWS

Although antecedents can be found in government reports and calls for conferences on cognitive issues in surveys as early as the late 1970s, the cognitive aspects of survey methodology (CASM) movement is generally thought to have begun with two conferences: the 1983 Advanced Research Seminar on Cognitive Aspects of Survey Methodology in the United States (CASM I, followed in 1997 by CASM II) and the 1984 Conference on Social Information Processing and Survey Methodology in Germany. These conferences were intended to create a new interdisciplinary field—bringing together cognitive psychologists, survey methodologists, and other social scientists—to apply cognitive science to survey research problems (Jabine et al., 1984; Tanur, 1999; Schwartz, 2007).

Three developments occurred at and just after CASM I that led to cognitive interviewing as the prevailing questionnaire pretesting technique in use today. One of these developments, according to Presser et al. (2004), was Loftus's influential work (1984) using the think-aloud method drawn from the field of psychology. The think-aloud method is a component of the cognitive interview and is discussed in more detail in the following sections. However, also detailed below, traditional survey methodology contributed a second part to cognitive interviewing— verbal probing.

A second development was the subsequent funding from the National Science Foundation (NSF) to support research to evaluate the technique and create the first cognitive laboratory at the National Center for Health Statistics (NCHS). Within a decade, three major U.S. government statistical agencies, as well as academic institutions and survey contract organizations, developed their own cognitive laboratories to begin using pretesting techniques like cognitive interviewing (Tanur, 1999; Presser et al., 2004).

A third CASM I component critical in the development of cognitive interviewing was Tourangeau's four-stage cognitive model (Tourangeau, 1984; also see Tourangeau et al., 2000). According to Willis (2005), while other cognitive models exist, much of the empirical work on cognitive factors that affect survey response take Tourangeau's four-stage (comprehension, retrieval, judgment, response) model into account. Figure 5.1 provides more detail on each stage of survey response. These four stages of survey response often form the foundation of cognitive interview protocols and analysis methods.

COGNITIVE INTERVIEWING CURRENT PRACTICE

Practitioners' Techniques

As Willis, DeMaio, and Harris-Kojetin (1999:135) note, there is "considerable inconsistency in the terminology and defining approach that characterize" cognitive interview protocols. Three broad approaches to cognitive interviewing that methodologists have advocated are think-alouds (Bolton & Bronkhorst, 1996), verbal probing, and an approach that mixes the two (Willis, 2005). These approaches are tied together under the umbrella of cognitive interviewing because they are all used to engage interviewers and participants in a one-on-one interaction in which participant verbalizations help the researcher identify where survey questions or other features do not work as intended (Conrad & Blair, 2004; Beatty & Willis, 2007).

Drawing on the work of Ericsson and Simon (1980), Loftus (1984) used a protocol that asked participants to think aloud as they provided their responses. Put simply, this think-aloud technique requires interview participants to describe their thoughts as they consider the question and their answer.

An alternative approach is to ask targeted questions about survey items or features (i.e., verbal probes). This approach grew out of the

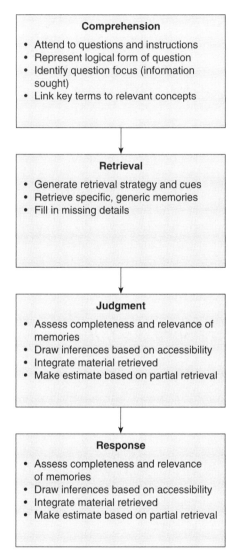

FIGURE 5.1 Tourangeau's four-stage cognitive model. *Source:* Tourangeau, Rips. & Rasinski, 2000, p. 8.

public opinion and survey research field (Willis, 2005). Verbal probes are used to elicit more detail about how participants process information when they review material or answer a survey item. Verbal probes are either concurrent or retrospective. Concurrent probes are asked after the survey item of interest and are intended to learn what the participant was thinking immediately after the processes were employed. The benefit to this type of verbal probe is that participants are more likely to recall what they were thinking as they processed the question and their answer. Retrospective probes are withheld until the end of the survey or some section of the survey, which allows for a more natural interview and reduces the chance of affecting future cognitive processes. Table 5.1 presents brief hypothetical examples of a think-aloud, concurrent probe, and retrospective probe.

TABLE 5.1
Hypothetical Examples of the Three Cognitive Interviewing Techniques

Think-alouds	
Interviewer	I'm going to read you a question. Please think out loud as you come up with your answer so I can better understand how you come up with it. Would you say that in general your health is . . . excellent, very good, good, fair, or poor?
Participant	Well, the last time I saw my doctor she told me I have high blood pressure and I was diagnosed as diabetic. She put me on a prescription for both and they are under control. But a friend of mine has severe rheumatoid arthritis that makes it nearly impossible for her to get around. I'm not as bad off as she is, so I'd say my health is fair.
Verbal (concurrent) probing	
Interviewer	Now thinking about your physical health, which includes physical illness and injury, for how many days during the past 30 days was your physical health not good?
Participant	5
Interviewer	How did you arrive at 5 days?
Participant	Well, I missed a week of work because I had a severe cold. But, come to think of it, I was sick the Sunday before the start of the week. I guess I should change my answer to 6.
Verbal (retrospective) probing	
Interviewer	Earlier in the survey I asked you, "Do you have one person you think of as your personal doctor or health care provider?"
Participant	Right, and I said "yes."
Interviewer	Right. What kinds of people do you think we are asking about when we ask this question?
Participant	I thought you were asking about regular MDs, RNs, chiropractors, and I even thought of my friend who sees a homeopathic doctor.

Cognitive Interviews in Practice: Present and Future

A set of common best practices has developed across a range of organizations over the past 30 years. This section summarizes those practices, followed by a description of how new technologies allow for an evolution of common practices to meet a changing (survey) world.

Recruiting: Cognitive interview participants are commonly recruited using nonprobability sampling techniques, such as advertisements placed in fliers or newspapers and contacting support groups (DeMaio & Rothgeb, 1996; Willis, 2005), through snowball or purposive sampling (Schechter et al., 1996; Irwin et al., 2009), through companies that maintain databases of study volunteers (Levin et al., 2009), through personal or professional networks (Nolin & Chandler, 1996), and by placing classified ads on websites like Craigslist.com (Murphy et al., 2007). Probability sampling is sometimes used for cognitive interviews (Caspar & Biemer, 1999), but this method of recruiting is relatively uncommon. Some recent research also suggests that a new method may have promise. For example, Head and others used advertisements on Facebook.com to recruit two special populations with modest success (Head et al., 2012; Head & Flanigan, 2012).

Sample size: Cognitive interviewing sample sizes tend to be relatively small. Willis (2005) has argued that relatively small sample sizes (5–15 per interviewing round) are adequate and that larger samples may provide diminishing returns. Driven by both U.S. Office of Management and Budget (OMB) regulations and cost considerations, smaller sample sizes seem to be the industry standard. However, many methodologists support the idea of incorporating larger sample sizes (Blair & Brick, 2009), and recent evidence suggests larger samples may be needed to identify a large proportion of existing problems, especially those that are rare (Blair & Conrad, 2011). Whether survey organizations incorporate large sample sizes into their studies in the future remains to be seen. At least three factors will likely play a role:

1. Further evidence—more empirical evidence is needed for federal agencies and large survey organizations to change existing standard operating procedures.
2. OMB regulations—for federally funded studies, the OMB needs to update regulations before federal agencies or organizations

that contract with them could conduct studies with larger samples.

3. Budgets—larger samples obviously require larger budgets. As of the writing of this chapter (in late 2012), federal agencies are either experiencing budget reductions or anticipating budget reductions. Strong evidence for larger samples would be needed to justify the cost increases that would be associated with such samples.

Mode: The most common mode for conducting cognitive interviews is in-person, although the location can be the laboratory, the participant's home, or some other location that is convenient for the participant (DeMaio & Rothgeb, 1996). Cognitive interviews have frequently been conducted in person because this method enables the interviewer to observe nonverbal cues that may indicate a problem (Willis, 2005). There are, however, good reasons to conduct cognitive interviews in other modes. Numerous survey modes are implemented for data collection. Cognitive interviews should be conducted to determine how well a questionnaire has been designed for the intended mode. For example, researchers have conducted cognitive interviews over the phone because the survey was ultimately going to be administered over the phone and doing so allowed for a test of how the questionnaire would perform in the mode of administration (Schechter & Beatty, 1994).

In-person cognitive interviewing can also be limited, especially in regard to geographic dispersion and rare populations. Traveling to different locations for each interview would be cost prohibitive; a more cost-effective strategy is to find another mode of contact. Methodologists have resolved this problem by conducting telephone interviews, but the telephone limits the interaction to verbal communication alone. New technologies, such as web-based videoconferencing and other online environments, may provide a middle ground in which at least some nonverbal cues can be observed, although perhaps not to the extent that can be observed in-person.

The emergence of web-based technologies such as social media, videoconferencing, and shared online spaces has set the stage for a period of innovation in cognitive interview applications and methodologies. The increased digital connectedness of individuals along substantive interests facilitates targeted recruitment of interview participants with desired demographics and/or other characteristics via online forums and social media. Furthermore, the growth of shared online spaces like Facebook, or virtual worlds like Second Life, along with web-based conferencing technology, provides a number of benefits to researchers. With video calling software such as Skype, researchers can conduct essentially face-to-face, private cognitive interviews with participants in any location, provided they have access to a computer, the Internet, and a webcam. Virtual worlds like Second Life offer this possibility as well, although both interviewer and participant use fully customized avatars to represent themselves rather than their real-life images via webcam. One key advantage of digital technologies for cognitive interviewing is potential access to a broad international population without the costs of travel. If, as the research of Blair and Brick (2009) and Blair and Conrad (2011) suggests, this lower cost access to more interview participants enables the routine expansion of cognitive interview sample sizes, researchers should be able to increase the number of errors identified in cognitive interviews, all at reduced cost. The pilot study discussed next incorporated Craigslist and Facebook for recruiting and compares cognitive interview results obtained through Second Life and Skype with in-person methods.

SECOND LIFE FOR SURVEY RESEARCH

Second Life (www.secondlife.com) is an online virtual world that allows users to self-represent through avatars they use to interact with others in the environment. The purposes of play include socializing, entertainment, and education. Previous research indicates that the online virtual world Second Life can be used for quick recruitment of users for interviews (Dean et al., 2012). Second Life is useful in recruiting targeted populations, including hard-to-reach populations like users with chronic medical conditions; gay, lesbian, and bisexual users; and users with sexually transmitted infections (Swicegood et al., 2012; Richards & Dean, 2012). Second Life is also a viable platform for data collection through virtual face-to-face and phone interviews, as well as through virtual-assisted self-interviewing or surveys administered inworld (Richards & Dean, 2012; Bell, Castronova, & Wagner, 2009; Head et al., 2012; Kamel Boulos & Toth-Cohen, 2009). Web surveys have also been successfully administered using Second Life (Chesney et al., 2009; Swicegood et al., 2012; Minocha et al., 2010; Foster, 2011). Through Second Life, researchers can recruit geographically diverse groups of individuals using nonprobability sampling techniques and collect data via various online collection modes.

The cognitive interviews for this study tested a questionnaire about Second Life users and their avatars. The questionnaire contained questions about physical appearance and behavioral characteristics of Second Life users' real-life selves as well as their primary avatars (e.g., questions about gender, race, hair color, personality traits, and disabilities). The questionnaire was designed to assess how similar users are to their avatars, as well as general use of web and social media content. The questionnaire also asked about health care as research has identified a number of health-related uses of Second Life (see Chapter 10 in this volume.) Because of the questionnaire's extensive Second Life content, testing it with people who use Second Life was critical.

METHODS

The purpose of this study was to conduct a pilot test of cognitive interviewing using both Second Life and Skype (www.skype.com), a voice-over-Internet-protocol (VoIP) software program that allows users to communicate through voice and chat. Interviews in Second Life were conducted between two personal avatars, one representing the interviewer and one the participant, simulating a face-to-face interaction in real time using voice chat with their real voices. The Skype interviews were conducted via webcam with videoconferencing, meaning that the interviewer and participant communicated essentially face-to-face. To determine the feasibility of using these two online modes compared with in-person cognitive interviews, we conducted a few in-person laboratory interviews offline.

Recruitment

Study participants were recruited through a paid Facebook ad and free ads placed on Craigslist.com. Through Craigslist we aimed to target Second Life users who were U.S. residents aged 18 years or older that could participate in an interview in one of our three modes: face-to-face, Second Life, or Skype. We placed ads in major metropolitan cities across the country including New York City, Los Angeles, Houston, and Minneapolis. In addition, participants were recruited in Research Triangle Park, NC, to conduct in-person cognitive interviews at RTI's headquarters. Other cities were chosen based on population density and regional variation. We placed ads in several West Coast cities to target Second Life users closer to the birthplace of Second Life—San Francisco. These locales included Santa Barbara, the San Francisco Bay area, and San Diego. We do not have evidence to support this assumption because usage statistics on Second Life are hard to find, but we suspected that, as a result of the heavy technology and computer industry focus of the economy on the West Coast, Second Life might have more users in those cities.

Unlike Craigslist, Facebook allows users to target specific populations within Facebook when posting ads. Through Facebook we targeted Second Life users based on the following criteria: (1) English speakers residing in the United States; (2) those that "like" Second Life on Facebook; and (3) those that "like" avatars in general. Each wave of recruitment targeted approximately 100,000 Facebook users. These ads were intended to be displayed on the main Facebook news feed pages for users who met these criteria. Because of ad suppression applications and various platforms of Facebook use (mobile, tablet, web), we could not verify that every member of the target audience saw the ad.

Three week-long waves of recruitment were conducted on both Craigslist and Facebook between March 16, 2012 and May 31, 2012. Each wave of recruitment highlighted one of two themes researchers developed to recruit Second Life users. In the first wave, researchers posted ads that emphasized a $40 Amazon gift card incentive. The second wave of recruitment was designed to target Second Life users who were interested in an interview because of their experiences in Second Life or their interest in Second Life research. The third wave of recruitment emphasized an incentive payment choice offered only in Wave 3, which is further described below.

Facebook was vastly more effective than Craigslist for recruiting our population. During the 3 weeks of active recruiting, we received 1,610 usable responses to our screener survey from Facebook and only 28 usable responses from Craigslist. (Usable meant that the person provided an avatar name or ID so we could follow up in Second Life to arrange an interview.) This result was surprising. Having previously used Craigslist to recruit (but not Facebook), we were not surprised by the 28 recruits from that venue, but we were not at all expecting 1,610 responses from Facebook. When scheduling the interviews, we ultimately found that we could not get responses from most of the Craigslist volunteers to arrange an interview time. Hence, we ended up drawing mostly from the vast pool of Facebook volunteers. Of the 40 interviews conducted for this study, 39 participants were recruited through Facebook and 1 through Craigslist. At this point, we can only speculate as to why Facebook was so much more effective. Perhaps the targeted ads are an untapped resource for cognitive interview recruitment or possibly Second Life users are a particularly active Facebook population—or Craigslist casts too broad of a net to be able to recruit specific but geographically dispersed populations like Second Life users. We hope to explore this question more with future trials of online recruitment techniques.

Screening

All ads included a link to a screener survey. People who encountered the ads on Craigslist or Facebook could copy and paste the URL into their personal web browser to complete this eligibility survey. In addition to determining eligibility, the screener collected data on several demographic indicators and, for the in-person sample, asked questions regarding geographic location to identify those who would be willing to drive to our facilities for an interview. Other questions were used to match participants to the two online modes, Second Life and Skype. For Skype, we selected only participants who reported that they currently had or were willing to download Skype. During sampling, researchers selected participants based on these criteria as well as demographic indicators such as gender, race, and age. Researchers also tried to select a sample balanced by gender, race, and age for each mode. When scheduling, researchers confirmed the availability of video capabilities for both Second Life and Skype interviews.

To recruit participants for in-person interviews, questions in the screener survey asked participants their state of residence. If participants lived in North Carolina, they were asked to specify their city and if they were willing to travel to RTI offices for an interview. All other participants had to be willing to conduct an interview in Second Life or via Skype. Participants who were not local to RTI offices were assigned to Second Life or Skype based on their video and voice capabilities and their availability.

Incentive

During the first two waves of recruitment, a $40 Amazon gift card was sent electronically to all study participants upon completion of the interview; in-person participants were given a printout that included the code to redeem their gift card online. For Second Life and Skype interviews, researchers typed the link where participants could redeem their gift card and their gift card code into a textbox. In the third and final wave of recruitment, an additional incentive option was added to determine whether Second Life users preferred to receive Linden dollars (the currency used within Second Life) or the equivalent amount as an Amazon gift card. During this wave, 17 out of 18 participants elected to receive their incentive in Linden dollars. For these 17 participants, the equivalent of $40 USD, approximately $10,000 Linden dollars, was transferred from the interviewer's account to each participant in Second Life upon completion of these interviews.

Think-Aloud and Probes

Interviewers used two approaches: think-aloud interviewing and verbal probing. At the onset of the interview, participants were asked to think aloud as they responded to each question, telling us everything that came to mind, regardless of whether it seemed important. Other questions specifically asked participants to think aloud as they answered the question. Concurrent verbal probes were also administered throughout the interview. At the end of the interview, all participants were asked a number of debriefing questions that included topics covered during the interview and questions regarding the type of computer they used for the interview. These questions were used to determine the type of computing technology required to participate in these interviews.

RESULTS

Overall Participant Characteristics

Of the 1,648 eligible screening participants, 40 participated in cognitive interviews. We planned 40 cognitive interviews from the beginning; the number of eligible screened participants was high because of the overwhelming efficiency of the Facebook ads. Table 5.2 shows the demographic characteristics of the participants interviewed. Of these

TABLE 5.2

Participant Demographics

	Number	Percentage
Gender		
Male	21	52.5%
Female	19	47.5%
Overall	40	100%
Age		
18–25	2	6.3%
26–35	13	40.6%
36–45	7	21.8%
46–55	7	21.8%
56–65	2	6.3%
66+	1	3.1%
Overall	32*	100%
Ethnicity		
Hispanic	5	12.5%
Not Hispanic	35	87.5%
Overall	40	100.0%
Race	Total	
White	19	47.5%
Black or African American	6	15.0%
American Indian or Alaska Native	10	25.0%
Asian	0	0.0%
Native Hawaiian or Pacific Islander	1	2.5%
Other	4	10.0%
Overall	40	100.0%
Education		
Less than high school	2	5.0%
Diploma or GED	7	17.5%
Some college	10	25.0%
Associate's degree	8	20.0%
Bachelor's degree	3	7.5%
Graduate degree	10	25.0%
Overall	40	100%
Household Income		
$20,000 or less	13	32.5%
$20,001–$40,000	14	35.0%
$40,001–$60,000	4	10.0%
$60,001–$80,000	4	10.0%
$80,001–$100,000	2	5.0%
More than $100,000	3	7.5%
Overall	40	100%

*Eight participants did not report their age.

participants, 21 (52.5%) were male. Sixty-two percent of participants were 26 to 45 years old. Ethnically, 12.5% of participants were of Hispanic origin. The racial makeup of the participant group was 47.5% white, 25% American Indian or Alaska Native, 15% African-American, 10% selected "other," and one participant (2.5% of sample) reported the racial category "Asian or Pacific Islander." Given that this was a convenience sample, we did not expect a representative group of participants by race, but 25% of cognitive interview participants for a particular study identified as American Indian—an unusually high number especially because no specific geographic or racial subpopulation was targeted. As a reference, the 2011 American Community Survey reported that 1.6% of the U.S. population is American Indian or Alaska Native (U.S. Census Bureau, 2012). Cognitive interview participants were well educated with 77.5% reporting some college or higher and 25% reporting having a graduate degree. Income tended to fall in the low-to-mid range, with 67.5% of the sample reporting $40,000 or less.

In addition to real-life demographic information, we were also interested in Skype use and mean age (years since creation) of Second Life avatar for each participant. Among cognitive interview participants, 82.5% had access to Skype prior to participating in the interview. The mean Second Life avatar age was 3.7 years among all cognitive interview participants.

Feasibility of Pilot Study

The pilot study showed that interviewing in Second Life and Skype was technically feasible. All participants were experienced Second Life users with access to the software, and most had access to Skype. Table 5.3 presents the overall video and audio quality of the Second Life and Skype interview recordings.[1] Among the two digital modes, most (30) were

---

[1] Of the 40 cognitive interviews, 5 were excluded from analysis for technical reasons related to the video recordings. Four out of 5 of these interviews were conducted in Second Life. These interviews were excluded because of audio failures caused by the recording software, a low-cost screen and audio-capture software purchased online. The recording for one interview, conducted face-to-face, was lost during the study, leaving 35 interviews available for analysis. Two of these were conducted in text chat because of technical difficulties or lack of voice capabilities in Second Life and Skype that were discovered during interviewing. Of the 35 interviews available for analysis, 17 were conducted in Second Life, 16 in Skype, and 2 in person.

TABLE 5.3
Audio and Video Quality of Second Life and Skype Interviews

Mode	Video Quality				Audio Quality			
	High	Medium	Low	Overall	High	Medium	Low	Overall
Second	17	0	0	17	7	7	3	17
Life	(100%)	(0.0%)	(0.0%)	(51.5%)	(41.2%)	(41.2%)	(17.7%)	(51.5%)
Skype	13	2	1	16	10	4	2	16
	(81.3%)	(12.5%)	(6.3%)	(48.5%)	(62.5%)	(25.0%)	(12.5%)	(48.5%)
Total	30	2	1	33	17	11	5	33
	(91.4%)	(6.1%)	(3.0%)	(100.0%)	(51.5%)	(33.3%)	(15.2%)	(100.0%)

identified as high video quality (all 17 of the Second Life interviews and 13 of the 16 Skype interviews). High video quality indicated no problems; medium quality indicated some problems, but video content (including facial expressions) was visible; and low indicated significant problems or interruptions, including failure of the video feed. Audio quality was more of a challenge. Seventeen interviews were coded as high audio quality (7 in Second Life and 10 in Skype); 11 as medium (7 in Second life and 4 in Skype); and 5 as low (3 in Second Life and 2 in Skype). High-quality audio was audio with no problems; medium-quality audio included some problems hearing the participant, but most audio content was understandable; and low-quality audio had major problems hearing content or audio failure. In interviews with audio failure, interviewers were usually able to complete the interview via the text chat function in Second Life and Skype. In the one interview with complete video failure in Skype, the interview was resumed at a later time when the video feed was functioning. Presumably the problem was a network connection.

Quality of Cognitive Interviews by Mode

Given that conducting the Second Life and Skype interviews was technically feasible, we examined the quality of the cognitive interview data collected across modes. Although cognitive interviews are designed to assess and remove measurement error in questionnaires, they have their own sources of error as well, as Conrad and Blair point out (2004). To understand the overall quality of Skype and Second Life cognitive interviews, we assessed the extent to which errors or problems occurred in each cognitive interview. Measures of cognitive interview mode quality included the number of times a cognitive interview participant was identified as "disengaged" with the interview, the number of nonverbal cues provided during the cognitive interview, and the total number of question problems identified. Furthermore, to obtain a more detailed understanding of the qualities of errors identified using each mode, we assessed type and severity of each problem.

To identify disengagement, nonverbal cues, and question problem incidence, type, and severity, the 35 cognitive interview video recordings were coded for errors. Coders viewed the video recordings and entered codes into a spreadsheet while viewing. To assure interrater reliability, each coder first reviewed the same 10 interviews. After reviewing and coding the first interview, coding results were tested with a Kappa test of reliability. The two coders achieved a 0.66 Kappa score, with $p < 0.05$. After interrater reliability was established, the remaining 25 interviews were split and coded separately between the two coders. Table 5.4 describes each code assigned.

Our assumptions used the in-person lab interviews as the gold standard. Because local recruitment challenges restricted us to conducting only three face-to-face interviews (two of which resulted in recordings), the analysis of face-to-face interviews is not statistically meaningful. We have included these in the results and the tables for comparison purposes, but the results should be interpreted as anecdotal.

Participant Disengagement

Table 5.5 displays participant disengagement in the cognitive interviews. Overall, incidence of participant disengagement was rare across modes. Participants seemed slightly more engaged in Second Life interviews than in Skype interviews. The 17 Second Life interviews yielded 14 observations of participant disengagement (about 0.8 observations per interview). The 16 Skype interviews yielded 18 observations of participant disengagement (about 1.1 observations per interview). The two in-person interviews revealed no observations of participant disengagement. The in-person interviews took place in a controlled laboratory environment, a quiet setting in an office building with no external interruptions, thus, minimizing opportunities for participant disengagement—although disruptions do occur in lab-based cognitive

TABLE 5.4
Quality Indicators of Cognitive Interview Video Recordings

Quality Indicator	Definition	Assumption
Participant Disengagement	Incidences of disengagement were coded if at any time the participant (1) temporarily left the interview (e.g., got up from computer and moved away from video feed); (2) asked the interviewer to hold on while they handled an event off-camera; (3) held a conversation with someone off-camera; (4) paused the interview, or otherwise failed to respond while attending to digital distractions (e-mail, chatting, online reading).	In-person interviews conducted in cognitive laboratory environments minimize interruptions and distractions for the participant. More interruptions are expected in digital formats, and the greater the number of interruptions, the lower the quality of cognitive interview data is assumed.
Nonverbal Cues	Incidences of nonverbal cues were coded if participants gave indications of misunderstanding or being confused about the question through facial expressions or body movements visible on camera.	In-person interviews allow researchers to observe physical behaviors in participants that indicate distress or trouble comprehending the question. These are most commonly seen in facial expressions but may also be seen in hand gestures or body language. Digital formats provide less opportunity to observe nonverbal cues. In Skype, depending on the camera position and video quality, nonverbal behaviors may be hard to make out. Avatars in Second Life make nonverbal gestures, but many of these are automated and out of the control of the participant. We assume that fewer nonverbal cues are associated with a decrease in cognitive interview data quality because less information is available to the researcher.

(Continued)

123

TABLE 5.4
(*Continued*)

124

Quality Indicator	Definition	Assumption
Number of Question Problems	This code identifies the total number of problems identified across all interviews in a given mode. The mean number of question problems is the average total number of problems identified per interview in each mode.	The purpose of a cognitive interview is to identify problems and potential problems with the questionnaire being tested. Therefore, we assume that the more problems identified in the cognitive interview, the higher the cognitive interview data quality.
Types of Question Problems	These codes identify the total number of each type of problem according to Tourangeau's cognitive response model (comprehension, retrieval, judgment, response) across all interviews in a given mode. The mean number of types of problems is the average total number of each type of problem identified per interview in each mode.	To get a more nuanced handle on the problems identified in the cognitive interview, instead of only coding a problem as a problem, problems were coded to each phase of Tourangeau's cognitive response process: comprehension, retrieval, judgment, and response. We wanted to know if certain modes were more likely better equipped to identify certain types of problems. The problems were identified with a specific stage of cognitive response, and the summation of types was used as the indicator for the total number of problems in the question.
Severity of Cognitive Problem	This code was marked each time a problem was identified as one in which the misinterpretation would result in a changed answer to the question.	When problems were identified, coders assessed whether the participant's misinterpretation of the question would result in a changed answer. Problems that result in a different answer are more severe and therefore more important types of problems to identify. The assumption is that the cognitive interview data quality is higher if a mode identifies more severe problems with the questionnaire.

TABLE 5.5
Participant Disengagement, by Mode

Mode	Total Number of Observations of Participant Disengagement	Mean Number of Observations of Participant Disengagement (SD)
Second Life (*N* = 17)	14	0.82 (1.67)
Skype (*N* = 16)	18	1.13 (1.75)
Face-to-Face2F (*N* = 2)	0	0 (0)
Total (*N* = 33)	32	0.91 (1.65)

interviews fairly regularly (cell phone calls, observations of activities going on outside the lab windows, or participants simply changing the subject in the cognitive interview.) On the other hand, the Skype and Second Life interviews took place in participants' homes, where any number of household, telephone, or computer distractions (including messages from other users of Skype and Second Life) competed against the interview for participant attention.

Nonverbal Cues

Data about nonverbal cues are presented in Table 5.6. The results for the nonverbal cue analysis met expectations. Skype interviews, where the participant's real facial features and mannerisms are visible, yielded approximately two nonverbal cues per interview, and Second Life interviews, where only nonverbal cues must be deliberately conveyed through the avatar, yielded less than one (0.5) nonverbal cue on average. In the two in-person interviews, one yielded 19 nonverbal cues, but the other yielded none.

TABLE 5.6
Nonverbal Cues, by Mode

Mode	Total Number of Nonverbal Cues	Mean Number of Nonverbal Cues (SD)
Second Life (*N* = 17)	9	0.53 (0.80)
Skype (*N* = 16)	34	2.13 (3.50)
Face-to-Face (*N* = 2)	19	9.50 (13.44)
Total (*N* = 35)	62	1.77 (3.92)

TABLE 5.7
Total Number of Problems, by Mode

Mode	Total Number of Problems Identified	Mean Number of Problems Identified	SD of Number of Problems Identified
Second Life ($N = 17$)	94	5.53	2.87
Skype ($N = 16$)	114	7.13	5.06
Face-to-Face ($N = 2$)	29	14.5	12.02
Total ($N = 35$)	237	6.677	4.88

Total Problems

Table 5.7 shows the total number of problems identified across all interviews in each mode were counted. In Second Life, 94 total problems were observed across 17 interviews, for an average of about 5.5 problems per interview. In Skype, 114 problems were observed across 16 interviews, for an average of about 7 problems per interview. In the two in-person interviews, one interview resulted in 6 problems and the other in 23 problems.

Type and Severity of Problems

Table 5.8 presents the total number of types of problems observed by mode, along with mean and standard deviation (mean; SD). The mean value is the average number of times a problem type was observed per interview. Overall, comprehension (2.46 per interview) and response (3.31 per interview) problems were observed far more frequently than retrieval (0.37 per interview) and judgment (0.46 per interview) problems. The limited number of observations restricts our ability to observe

TABLE 5.8
Number and Type of Problem, by Mode (Mean, Standard Deviation)

Mode	Comprehension	Retrieval	Judgment	Response
Second Life ($N = 17$)	30 (1.76; 1.25)	8 (0.47; 0.62)	5 (0.29; 0.59)	42 (2.47; 1.73)
Skype ($N = 16$)	43 (2.69; 1.92)	4 (0.25; 0.58)	8 (0.50; 0.89)	66 (4.13; 2.31)
Face-to-Face ($N = 2$)	13 (6.50; 4.94)	1 (0.50; 0.71)	3 (1.50; 2.12)	8 (4.00; 2.83)
Total ($N = 35$)	86 (2.46; 2.08)	13 (0.37; 0.60)	16 (0.46; 0.85)	116 (3.31; 2.17)

a high incidence of these fairly low frequency problems, although for comprehension and response, more problems seem to be observed in Skype interviews than in Second Life. In the two face-to-face interviews, the first interview (with 23 total problems) had 10 comprehension, 3 judgment, 1 retrieval, and 6 response problems. The second interview (with only 6 total problems) had 3 comprehension and 2 response problems.

Furthermore, we rarely observed a problem so severe that continued misunderstanding would have changed the participant's answer to a question. Overall, only 14 questions out of 82 had a problem that would have resulted in a changed answer in at least one interview. Two of these were observed in Second Life, 8 in Skype, and 4 in the in-person interviews.

CONCLUSIONS

Despite the mediated distance between interviewer and participant introduced by virtual communication technology, cognitive interviewing is feasible in both Second Life and Skype. We were able to recruit and schedule interviews with a nationally dispersed population of Second Life users, and interviews took place with minimal technical difficulty, despite the loss of several video recordings. (For future research, we recommend choosing a high-quality and well-known screen capture recording software to eliminate the problem of inferior or unusable recordings.)

We continue to assume that in-person interviewing is superior to virtual cognitive interviewing, although anecdotal evidence from two interviews does not prove or disprove this assumption. Cognitive interviewing is a technique designed to uncover measurement errors in surveys before they are fielded. At minimum, the greatly reduced distractions observed in in-person interviews may contribute to more errors discovered. Yet, perhaps virtual cognitive interviews conducted in settings more like real life, with its interruptions and distractions, would more effectively identify errors stemming from survey participants' inattention. The face-to-face comparison group does not offer enough observations to be able to tell for sure. In those three interviews, participants seemed to be less disengaged from the process, more nonverbal cues were available to the interviewer and analyst, and a higher mean number of problems were identified. Yet, the results for face-to-face interviews are confounded by the fact that participants who participated in the in-person interviews for this study were extremely enthusiastic volunteers. That is, as we were pilot testing the ability to recruit subjects from the U.S. population, the lab participants had to be willing (and able) to drive to RTI headquarters to complete the cognitive interviews. Therefore, the lab volunteers may have been more motivated participants in general, thus, generating better quality data.

As expected, Skype interviews produced more nuanced findings than the results obtained from Second Life interviews. That is, the ability to see the participant's real face, rather than their Second Life avatar, resulted in more observations of potential measurement errors in the form of question problems as well as nonverbal cues. However, Second Life interviews were slightly less likely to indicate participant disengagement. These results could have occurred for two possible reasons: (1) As a virtual world, Second Life is an immersive experience that may lend itself to more user focus; (2) on the other hand, participant disengagement may be harder to measure without being able to view the participant's real face and facial features. On average, Second Life results, although usable, seem to provide less information to analysts.

DISCUSSION AND FUTURE RESEARCH

We predict that long-distance virtual cognitive interviewing, be it in Second Life, Skype, or a yet-to-be-identified platform, will grow over time to be a valuable tool to survey designers. Despite the observation in this study that in-person interviews seem to yield more observations of questionnaire problems, virtual cognitive interviewing should be evaluated according to costs and benefits. For this study, only 3 of the 40 interviews were conducted with subjects in North Carolina, where the researchers were located. Interviewing these same 40 participants in real life would have meant extensive travel costs. This study was not designed as a cost–benefit analysis, but if conducting 40 cognitive interviews via Skype is about half the cost of conducting 10 interviews in real life, the additional observations obtained at cheaper cost may be worth a lower per-interview number of problem observations. Furthermore, web camera and software technology continue to improve. Perhaps with a few years' worth of enhancements, such as higher quality video capture and faster video processing speeds, Skype interviews, or even Second Life interviews, will become qualitatively equivalent to in-person interviews. Future research should continue to test these platforms with greater controls (for example, random assignment to mode) and with larger sample sizes. One potential advantage of Second Life for cognitive interviewing in particular is the added layer of privacy that response via avatar may add to the interview process. Particularly for cognitive interviews about very sensitive topics, participants may be more forthcoming and more comfortable responding to questions wearing the masks of their avatars. Designers of questionnaires about very sensitive topics should consider the use of Second Life for at least one round of cognitive interviewing. Regardless of mode, virtual cognitive interviewing offers a valuable resource to survey designers.

REFERENCES

BEATTY, P. (2002). Cognitive interview evaluation of the Blood Donor History Screening Questionnaire. *Final Report of the AABB Task Force to Redesign the Blood Donor Screening Questionnaire.* Prepared for the U.S. Food and Drug Administration.

BEATTY, P. C., & WILLIS, G. B. (2007). Research synthesis: The practice of cognitive interviewing. *Public Opinion Quarterly, 72,* 287–311.

BELL, M. W., CASTRONOVA, E., & WAGNER, G. G. (2009). Surveying the virtual world: A large-scale survey in Second Life using the Virtual Data Collection Interface (VDCI). *German Council for Social and Economic Data (RatSWD) Research Notes, 40,* 1–49.

BIEMER, P., & LYBERG, L. (2003). *Introduction to survey quality.* New York: Wiley.

BLAIR, J., & BRICK, P. D. (2009). *Current practices in cognitive interviewing.* Presented at the annual conference of the American Association for Public Opinion Research (AAPOR), Hollywood, FL.

BLAIR, J., & CONRAD, F. G. (2011). Sample size for cognitive interview pretesting. *Public Opinion Quarterly, 75,* 636–658.

BOLTON, R. N., & BRONKHORST, T. M. (1996). Questionnaire pretesting: Computer-assisted coding of concurrent protocols. In N. Schwarz & S. Sudman (Eds.), *Answering questions: Methodology for determining cognitive and communicative processes in survey research* (pp. 37–64). San Francisco: Jossey-Bass.

CARBONE, E. T., CAMBELL, M. K., & HONESS-MOREALE, L. (2002). Use of cognitive interview techniques in the development of nutrition surveys and interactive nutrition messages for low-income populations. *Journal of the American Dietetic Association, 102,* 690–696.

CASPAR, R., & BIEMER, P. P. (1999). The use of cognitive laboratory interviews for estimating production survey costs and respondent burden. *Proceedings of the Section on Survey Methods Research, American Statistical Association,* 192–197.

CHESNEY, T., CHUAH, S. H., & HOFFMANN, R. (2009). Virtual world experimentation: An exploratory study. *Journal of Economic Behavior & Organization, 72,* 618–635.

CONRAD, F. C., & BLAIR, J. (2004). Data quality in cognitive interviews: The case of verbal reports. In S. Presser, M. P. Couper, J. T. Lessler, E. Martin, J. Martin, J. M. Rothgeb, & E. Singer (Eds.), *Methods for testing and evaluating survey questions* (pp. 67–87). New York: Wiley.

DEAN, E. F., COOK, S. L., MURPHY, J. J., & KEATING, M. D. (2012). The effectiveness of survey recruitment methods in second life. *Social Science Computer Review, 30,* 324–338.

DEMAIO, T. J., & ROTHGEB, J. M. (1996). Cognitive interviewing techniques: In the lab and in the field. In N. Schwarz & S. Sudman (Eds.), *Answering questions: Methodology for determining cognitive and communicative processes in survey research* (pp. 177–195). San Francisco, CA: Jossey-Bass.

ERICSSON, K., & SIMON, H. (1980). Verbal reports as data. *Psychological Review, 87,* 215–251.

Federal Committee on Statistical Methodology (FCSM). (2001). *Measuring and reporting sources of error in surveys.* Statistical Working Paper 31. Retrieved from http://www.fcsm.gov/01papers/SPWP31_final.pdf.

FOSTER, K. N. (2011). *The second life of social science research: An assessment of sampling and data collection methods, data quality, and identity construction in virtual environments.* Ph.D. Dissertation. University of Georgia, Athens, GA.

GROVES, R., & LYBERG, L. (2010). Total survey error: Past, present, and future. *Public Opinion Quarterly, 74,* 849–879.

GROVES, R., FLOYD, F., COUPER, M., SINGER, E., & TOURANGEAU, R. (2004.) *Survey methodology.* New York: Wiley.

HARRIS-KOJETIN, L. D., FOWLER, JR. F. T., BROWN, J. A., SCHNAIER, J. A., & SWEENY, S. F. (1999). The use of cognitive testing to develop and evaluate CAHPS™ 1.0 core survey items. *Medical Care, 37,* MS10–MS21.

HEAD, B. F., DEAN, E. F., KEATING, M. D., SWICEGOOD, J. E., POWELL, R., & SAGE, A. J. (2012). *Recruiting virtual world users for cognitive interviews: A comparison of Facebook and Craigslist.com advertisements.* Presented to the Southern Association for Public Opinion Research, Raleigh, NC.

HEAD, B. F., & FLANIGAN, T. S. (2012). *Recruiting older adults for cognitive interviews: Comparing traditional and emerging techniques.* Presented to the Southern Association for Public Opinion Research, Raleigh, NC.

IRWIN, D. E., VARNI, J. W., YEATTS, K., & DEWALT, D. A. (2009). Cognitive interviewing methodology in the development of a pediatric item bank: A patient reported outcomes measurement information system (PROMIS) study. *Health and Quality of Life Outcomes, 7,* 1–10.

JABINE, T. B., STRAF, M. L., TANUR, J. M., & TOURANGEAU, R. (1984). *Cognitive aspects of survey methodology: Building a bridge between disciplines.* Washington, DC: National Academy Press.

KAMEL BOULOS, M. N., & TOTH-COHEN, S. (2009). The University of Plymouth sexual health sim in Second Life: Evaluation and reflections after 1 year. *Health Information and Libraries Journal, 26,* 279–288.

LEVIN, K., WILLIS, G. B., FORSYTH, B. H., KUDELA, M. S., STARK, D., & THOMPSON, F. E. (2009). Using cognitive interviews to evaluate the Spanish-language translation of a dietary questionnaire. *Survey Research Methods, 3,* 13–25.

LOFTUS, E. (1984). Protocol analysis of responses to survey recall questions. In T. Jabine, J. Straf, & R. Tourangeau (Eds.), *Cognitive aspects of survey methodology: Building a bridge between disciplines* (pp. 61–64). Washington, DC: National Academy Press.

MILLER, K. (2003). Conducting cognitive interviews to understand question-response limitations. *American Journal of Health Behavior, 27,* S264–S272.

MINOCHA, S., TRAN, M. Q., & REEVES, A. J. (2010). Conducting empirical research in virtual worlds: Experiences from two projects in Second Life. *Journal of Virtual Worlds Research, 3,* 1–21.

MURPHY, J. J., SHA, M., FLANIGAN, T. S., DEAN, E. F., MORTON, J. E., SNODGRASS, J. A., & RUPPENKAMP, J. W. (2007). *Using Craigslist to recruit cognitive interview respondents.* Presented to the Midwest Association for Public Opinion Research, Chicago, IL.

NOLIN, M. J., & CHANDLER, K. (1996). *Use of cognitive laboratories and recorded interviews in the National Household Education Survey* (NCES 96-332). Washington, DC: U.S. Department of Education. National Center for Education Statistics.

PRESSER, S., COUPER, M. P., LESSLER, J. T., MARTIN, E., MARTIN, J., ROTHGEB, J. M., & SINGER, E. (2004). Methods for testing and evaluating survey questions. In S. Presser, M. P. Couper, J. T. Lessler, E. Martin, J. Martin, J. M. Rothgeb, & E. Singer (Eds.), *Methods for testing and evaluating survey questions* (pp. 1–22). New York: Wiley.

REDLINE, C., SMILEY, R., LEE, M., & DEMAIO, T. (1998). *Beyond concurrent interviews: An evaluation of cognitive interviewing techniques for self-administered questionnaires.* Working Papers in Survey Methodology, No. 98/06. Washington, DC. Retrieved from http://www.census.gov/srd/papers/pdf/sm98-06.pdf.

RICHARDS, A. K., & DEAN, E. F. (2012). *Gaming the system: Inaccurate responses to randomized response technique items.* Presented to the American Association for Public Opinion Research Annual Conference, Orlando, FL.

SCHECHTER, S., & BEATTY, P. (1994). *Conducting cognitive laboratory tests by telephone* (Cognitive Methods Staff Working Paper No. 8). Hyattsville, MD: Centers for Disease Control and Prevention/National Center for Health Statistics.

SCHECHTER, S., BLAIR, J., & VANDE HEY, J. (1996). Conducting cognitive interviews to test self-administered and telephone surveys: Which methods should we use?

Proceedings of the Section on Survey Methods Research, American Statistical Association, 1–17.

SCHWARZ, N. (2007). Cognitive aspects of survey methodology. *Applied Cognitive Psychology, 21*, 277–287.

SWICEGOOD, J. E., HAQUE, S. N., DEAN, E. F., RICHARDS, A. K., & HEAD, B. F. (2012). *Recruiting special and hard-to-reach populations in Second Life study: Recruiting through the eyes of an avatar.* Presented at 140th annual meeting of the American Public Health Association, San Francisco, CA.

TANUR, J. M. (1999). Looking backwards and forwards at the CASM movement. In M. G. Sirken, D. J. Herrman, S. Schechter, N. Schwarz, J.M. Tanur, & R. Tourangeau (Eds.), *Cognition and survey research* (pp. 13–19). New York: Wiley.

TOURANGEAU, R. (1984). Cognitive science and survey methods: A cognitive perspective. In T. Jabine, J. Straf, & R. Tourangeau (Eds.), *Cognitive aspects of survey methodology: Building a bridge between disciplines* (pp. 73–100). Washington, DC: National Academy Press.

TOURANGEAU, R., RIPS, L. J., & RASINSKI, K. (2000). *The psychology of survey response.* Cambridge, UK: Cambridge University Press.

U.S. Census Bureau (2012). *American Indian and Alaska Native Heritage Month: November 2012.* News Release. Retrieved February 17, 2013, from http://www.census.gov/newsroom/releases/archives/facts_for_features_special_editions/cb12-ff22.html.

WILLIS, G. B. (2005). *Cognitive interviewing: A tool for improving questionnaire design.* Thousand Oaks, CA: Sage.

WILLIS, G., DEMAIO, T., & HARRIS-KOJETIN, B. (1999). Is the bandwagon headed to the methodological promised land? Evaluating the validity of cognitive interviewing techniques. In M. G. Sirken, D. J. Herrman, S. Schechter, N. Schwarz, J. M. Tanur, & R. Tourangeau (Eds.), *Cognition and survey research* (pp. 133–153). New York: Wiley.

ZUCKERBERG, A., & LEE, M. (1997). *Better formatting for lower response burden.* Working Papers in Survey Methodology, No 97/02. Washington, DC. Retrieved from http://www.census.gov/srd/papers/pdf/sm97-2.pdf.

CHAPTER 6

Second Life as a Survey Lab: Exploring the Randomized Response Technique in a Virtual Setting

Ashley Richards and Elizabeth Dean, *RTI International*

Second Life is an online virtual world that provides a three-dimensional (3-D) graphical, interactive environment. This environment represents (and often expands on) real life. Because Second Life content is completely user-generated, researchers can design and manipulate environments and avatars to test social interactions with a consistency that is not possible in the real world. This chapter describes the benefits of conducting research in Second Life and reports findings from the authors' own research in which we used Second Life to evaluate a method of asking survey questions, the randomized response technique (RRT).

Social Media, Sociality, and Survey Research, First Edition.
Edited by Craig A. Hill, Elizabeth Dean, and Joe Murphy.
© 2014 John Wiley & Sons, Inc. Published 2014 by John Wiley & Sons, Inc.

OVERVIEW OF SECOND LIFE

Second Life (http://secondlife.com) is an online 3-D virtual world used internationally for gaming, socializing, and creating art. A significant component of this virtual world is its economy, in which users buy and sell virtual goods and services. To support this economy, Second Life maintains its own currency, the Linden dollar, which exchanges with the U.S. dollar at a rate of about 250 Lindens to 1 U.S. dollar, as of December 8, 2012 (Shepherd, 2012). In Second Life, users represent themselves with highly customized avatars, or digital self-representations, which vary from quasirealistic presentations of the users' real-life characteristics to fantastical creatures, robots, vampires, or superheroes. In avatar form, users walk, run, fly, and teleport around the Second Life world, engaging in activities for business and for play. In this sense, Second Life is like a massive multiplayer online game although the world is more free-form than an overarching game world (such as *World of Warcraft* or *Minecraft*). That is, users, known as residents of Second Life, can choose to participate in inworld games or just engage in other activities.

Linden Lab launched Second Life in 2003, and in 2009, Second Life had over 15 million accounts (unique users) (Dean et al., 2012). Official usage statistics were last reported by Linden Lab in October 2011. At that time, about 1 million unique users logged in to Second Life on average each month and about 50,000 users were logged in at any given moment (Au, 2011). In April 2012, Second Life was listed as tenth on Nielsen's top ten list of PC games (Au, 2012).

RESEARCH IN SECOND LIFE

Fitting into both the conversational and community levels of the sociality hierarchy, Second Life and other online virtual worlds represent the real world while facilitating virtual—but still authentic—interactions among people. As a result, virtual worlds, including Second Life, have been considered a promising resource for social scientists (Bainbridge, 2007; Castronova & Falk, 2008; Harrison et al., 2011). More recently, Second Life has been suggested as a resource for engaging with hard-to-reach populations, and particularly those with specific health conditions (see Chapter 10 of this volume).

Because of its active economy, Second Life has attracted market researchers who conduct surveys and focus groups inworld. One company, the Social Research Foundation, manages the First Opinions Panel, a panel of Second Life users who are recruited and interviewed similarly to real-life market research panels (see http://www. socialresearchfoundation.org/). Second Life is a useful environment for conducting quick-turnaround, cost-effective surveys. Although the unknown population parameters such as the precise number of users and their demographic characteristics make it difficult to draw a representative sample of Second Life users, convenience samples are accessible in Second Life for conducting self-administered surveys at inworld automated kiosks, or for conducting interviewer-administered surveys in which an avatar interviewer surveys an avatar respondent (Dean et al., 2012). The mode holds great potential for questionnaire pretesting methods and for collecting qualitative data using focus groups, cognitive interviews, and in-depth or ethnographic interviews. In addition to its accessibility for conducting quick-turnaround surveys of convenience samples, another benefit of conducting surveys in Second Life is its ability to act as a survey laboratory.

The field of survey methodology is based on empirical findings generated through experimental research. Much of our knowledge of survey design features comes from experiments that manipulate particular features, such as the color and layout of paper questionnaires, the timing and delivery type of prenotification mailings, or the size of answer boxes in web surveys. Some factors relating to the survey interviewer and survey environment are more difficult to control and to study. For example, the race of the interviewer has been found to influence survey responses to questions about racial attitudes. Krysan and Couper (2003) found that African Americans report less liberal opinions to white interviewers. Through a series of framing experiments in a public opinion survey, Davis and Silver (2003) concluded that African American respondents answered fewer political knowledge questions correctly when interviewed by a white interviewer than when interviewed by an African American interviewer. Similarly, the gender of the interviewer has been shown to affect responses to questions about feminism in interviewer-administered surveys (Kane & Macauley, 1993), web surveys (gender conveyed by interviewer image), and interactive voice response phone surveys (gender conveyed by interviewer voice) (Tourangeau et al., 2003).

Lab-based research in Second Life enables precise manipulation of factors related to the social interaction between interviewer and respondent, as well as the environment in which surveys are administered, which are difficult to manipulate in real life. For example, in an (avatar) interviewer-administered survey of 60 Second Life residents, Murphy et al. (2010) manipulated only the body mass of the interviewer and found that respondents who were interviewed by the heavier interviewer reported less frequent real-life exercise and higher real-life body mass indexes compared with respondents assigned to the thinner interviewer.

The study described in this chapter examines a different survey feature, independent of interviewer characteristics. It takes advantage of the highly customizable virtual world environment to provide a means for assessing the RRT, a tool for allowing a more private response to sensitive survey questions.

THE RANDOMIZED RESPONSE TECHNIQUE

The RRT is used to encourage accurate responding to sensitive survey questions. When using the RRT, respondents are given two questions (one sensitive and one nonsensitive with a known response distribution) and are instructed to answer one of them. The question to be answered is determined by the outcome of a random act with a known probability (e.g., a coin toss), which only the respondent sees. Researchers do not know which question each respondent answered, but they can estimate proportions for each response to the sensitive question.

Findings about the RRT's effectiveness have been mixed, but a meta-analysis of RRT validation studies concluded that the RRT obtains more accurate estimates than direct questioning (Lensvelt-Mulders et al., 2005). The RRT has been found to elicit higher reporting of sensitive behaviors across a variety of substantive areas, including levels of induced abortion (Lara et al., 2004), misrepresentation on job applications (Donovan et al., 2003), employee theft (Wimbush & Dalton, 1997), illegal drug use (Zdep et al., 1979), and child abuse (Zdep & Rhodes, 1977). Researchers have typically assumed that if the RRT results in higher reporting of sensitive behaviors than direct questioning, then the RRT reduces measurement error because sensitive behaviors tend to be underreported.

However, if respondents do not implement the RRT correctly, it may not actually reduce error as intended. Coutts and Jann (2011), among others, have suggested that respondents do not comply with the RRT because they do not understand how to follow the procedure or because they do not understand how it protects their anonymity. Respondents also may not comply with the RRT because they do not want to appear to have endorsed a particular response.

Evaluating the true effectiveness of the RRT is difficult because researchers are blind to the outcome of the randomizing device. Even if the outcome *is* known, it is impossible to observe the cognitive processes occurring in respondents' minds. Because it is impossible to confirm that the procedure was followed, researchers have typically assumed that higher reporting of undesirable attributes signals the RRT's success. However, RRT implementation errors may lead to an increase in reporting of *all* attributes, not just undesirable attributes (Holbrook & Krosnick, 2010). Second Life is a beneficial setting for conducting RRT investigations because, unlike real life, the setting provides a unique form of omniscience. In this study, we took advantage of the Second Life setting to control the RRT randomizing device.

Study Design

We conducted an exploratory study to examine respondents' understanding of the RRT. First, we wanted to learn whether respondents truly understand how to follow the RRT procedure. Second, we aimed to distinguish between intentional and unintentional reporting errors when using the RRT. Intentional reporting errors may be a result of socially desirable responding, while unintentional errors may be a result of the cognitive demands of the technique.

The Second Life setting of this study was a critical part of its design. We studied the accuracy of respondent reporting in two ways. First, we controlled the outcome of the randomizer (a coin toss) so we knew which question respondents should answer.[1] The Second Life setting enabled such control. Second, respondents should have been able to answer the nonsensitive question in each pair, allowing us to assume that incorrect responses to nonsensitive questions were a result of other factors, such as implementation errors.

---

[1] Second Life enables users to create, design, and interact with objects in Second Life. We designed the coin so we could control the outcome of each coin toss, unbeknownst to the respondents.

FIGURE 6.1   Nonsensitive question prompt example.

The nonsensitive questions took one of two forms. The first form was not actually a question but an instruction to "Respond Yes" or "Respond No." The second form showed respondents an image and asked them a basic question about it. For example, the following image, which displays a blue triangle, a yellow star, and a red rectangle, was provided with the nonsensitive question, "Is the triangle red?" (see Figure 6.1; color version not available).

The study had two components: to assess comprehension of the RRT and compliance with the procedures. We first investigated whether respondents were capable of adhering to RRT procedures by pairing two innocuous questions. For example, we showed images of a tree and a white cat along with the questions "Does the cat have orange fur?" and "Is the cat larger than the tree?" (see Figure 6.2; color version not available).

We then paired sensitive and innocuous questions to examine the influence of social desirability on accuracy of responses. Several items

FIGURE 6.2   Innocuous question prompt example.

directed respondents to answer an innocuous question in a manner that would be undesirable if that response were given for the sensitive question in the pair. For example, we paired the sensitive[2] question "Did you vote in the presidential election held on November 4, 2008?" with the innocuous question "Is the triangle red?" and the corresponding image of colored shapes. Because of the outcome of the coin toss, respondents were supposed to answer the innocuous (triangle) question with a "No" response. However, a "No" response to the voting question would be socially undesirable.

In comparison, other innocuous questions had answers that would be desirable if given for the sensitive question. An example of this pairing was a sensitive question asking if the respondent ever had a sexually transmitted disease and an innocuous question, accompanied by an image of a small blue square and a large red square, asking if the blue square was larger than the red square. Because the answer to the innocuous question is "No," the answer would be desirable if provided for the corresponding sensitive question.

To test our research questions, we drew a convenience sample of Second Life residents ages 18 and older living in the United States. We recruited respondents through word of mouth in Second Life and by posting advertisements/notifications in the Second Life classifieds, Second Life forums, Second Life Facebook pages, and Craigslist. The ads stated that we were looking for Second Life residents to complete a survey in Second Life and that respondents would receive 500 Linden dollars[3] for participating. Of the 352 respondents who completed the screening survey to express interest in participating, 108 were eligible and were invited to participate in the study. Of the 76 respondents who replied to the invitation, all but one agreed to participate, for a total sample size of 75 and a participation rate[4] of 69%.

---

[2] A survey question is sensitive if it asks respondents to admit to having violated a social norm. Most survey questions about voting are sensitive because voting is a socially desirable behavior and reporting that one did not vote is undesirable. Questions about voting typically have higher rates of item nonresponse compared with other questions (Tourangeau & Yan, 2007).

[3] Linden Dollars are the currency used in Second Life. At the time of the interviews, L$500 was valued at approximately US$2.

[4] As suggested by the American Association for Public Opinion Research, the participation rate is used in lieu of a response rate because the sample was a nonprobability sample (American Association for Public Opinion Research Cell Phone Task Force, 2010).

TABLE 6.1

Respondent Demographics by Interview Mode

	Face-to-Face ($n = 24$)	Voice ($n = 20$)	Web ($n = 31$)	TOTAL ($N = 75$)
Sex				
Male	38%	50%	29%	37%
Female	63%	50%	71%	63%
Age				
18–24	25%	30%	23%	25%
25–40	58%	35%	52%	49%
41–60	17%	25%	26%	23%
61+	0%	10%	0%	3%
*Race**				
White	92%	80%	90%	88%
Black or African American	4%	5%	6%	5%
American Indian/Alaska Native	4%	10%	3%	5%
Asian	0%	0%	3%	1%
Other	0%	15%	0%	4%
Did not specify	0%	5%	0%	1%
Education				
High school or less	29%	25%	26%	27%
Some college (no Bachelor's)	54%	50%	48%	51%
Bachelor's	13%	15%	13%	13%
Graduate	4%	5%	10%	7%
Did not specify	0%	5%	3%	3%
Employment				
Full-time	29%	20%	35%	29%
Part-time	13%	10%	16%	13%
Homemaker	21%	10%	6%	12%
Retired	4%	25%	3%	9%
Student	8%	10%	16%	12%
Unemployed	25%	20%	19%	21%
Did not specify	0%	5%	3%	3%

*Could select more than one response

Respondents were assigned to one of three modes: face-to-face interview in Second Life, voice chat interview in Second Life, or web. When assigning modes we considered interviewer availability and respondent demographics. We aimed to achieve roughly equal demographic distributions across modes in case RRT responding was related to demographic characteristics. See Table 6.1 for respondent demographics across modes. In-person and voice interviews were conducted at RTI

FIGURE 6.3   In-person interview configuration.

International's virtual office in Second Life. The web mode differed from the in-person and voice modes in that respondents were sent a link to the survey that they completed in their web browser.

For both the in-person and voice modes, the interviewer and respondent communicated through the voice chat feature in Second Life, enabling them to talk aloud to each other. Respondents were given a coin to flip during the survey. Before asking each question, the interviewer set the coin to determine whether it would result in heads or tails. Respondents could not see that the interviewer had control over the coin. In the (virtual) in-person mode, the respondent sat across a table from the interviewer. The coin was on the table in front of the respondent, and a divider was placed on the table to make it appear that the coin flip was private (see Figure 6.3). In the voice mode, the interviewer and respondent were in different rooms and could not see each other, but they could talk to each other. This mode was meant to resemble a telephone interview. To make it even more like a phone interview, the respondent and interviewer never saw each other's avatars.

The setup was slightly different for web respondents, who watched a video of a coin flipping rather than flipping it themselves. This procedure was used to ensure the coin was synched properly with the web survey. The coin toss video may have reduced some respondents' trust in the method, so we took steps to reduce skepticism about whether the web survey coin toss was truly random, such as omitting a back button from the survey and ensuring that the coin did not reflip if the page was refreshed.

RESULTS

First, we examined whether respondents understood the RRT procedure by presenting them with pairs of innocuous questions with obvious, and opposite, answers. We found that 12% of respondents answered one of these two "understanding items" incorrectly, which was significant, $t(74) = -3.18$, $p < 0.01$, 95% confidence interval (CI) [0.902, 0.978]. This percentage is higher than we expected, considering that respondents were guided through an example and those in the face-to-face and voice modes were allowed to ask questions about the procedure before beginning these questions. Because they did not demonstrate their understanding of the procedure, we omitted the 12% who answered incorrectly from the remaining analyses. The demographics of this subset of respondents were not significantly different from the respondents who answered the understanding items correctly. However, a disproportional number of respondents who answered the items incorrectly were using the web mode (67%), so the mode may have possibly influenced their understanding.

Mixed throughout the rest of the survey were pairs of questions in which one question was sensitive and the other innocuous. In "undesirable response" items, the coin toss directed respondents to answer the innocuous question in a way that would be undesirable if provided for the sensitive question in the pair. The percentage of respondents answering incorrectly was significant or marginally significant for each of these undesirable response items. Across all undesirable response items, the percentage each respondent answered correctly was significantly lower than expected, $t(65) = -3.72$, $p < 0.001$, 95% CI [0.895, 0.968].

The survey also included "desirable response" items, pairing a sensitive and an innocuous question, but unlike the undesirable response items, in these items, the coin toss directed respondents to provide a response to the innocuous question that would be desirable if given for the sensitive item. The percentage of respondents answering these items correctly was also lower than expected, $t(65) = -1.76$, $p < 0.05$, 95% CI [0.951, 1.003].

TABLE 6.2
Belief That Coin Toss Outcome Was Private

	In-Person ($n = 24$)	Voice ($n = 20$)	Web ($n = 31$)
Yes	50%	50%	23%
No	33%	35%	77%
Don't know*	17%	15%	—

$p < 0.01$.
*An explicit "Don't know" response was not offered, but "Don't know" responses were accepted in the interviewer-administered modes.

We then conducted a paired t-test to compare each respondent's percentage of incorrect items on desirable and undesirable response items. Respondents answered significantly more desirable response items correctly than undesirable response items, $t(65) = 2.05$, $p < .05$, 95% CI [0.895, 0.968]. This finding suggests that respondents were only selectively following the procedure.

Overall, we found that only 66.7% of respondents followed the procedure for every RRT item, which was also significantly lower than expected, $t(74) = -5.16$, $p < 0.001$, 95% CI [0.915, 0.962]. The greatest rate of procedural noncompliance on an item was 13.3%. No significant differences were found in RRT compliance by demographic characteristics or survey mode.

At the end of the survey, we asked respondents a series of debriefing questions to assess their thoughts about the interview. One question was, "During the survey, were you convinced that no one else would know whether you flipped heads or tails?"[5] As shown in Table 6.2, responses differed significantly across modes, $\chi 2$ $(4, N = 75) = 15.1$, $p < 0.01$. Web respondents were less likely to believe that the outcome of the coin toss was private.

Next we compared the accuracy of responses to the RRT items for respondents who did and did not believe the coin toss was private.[6] We predicted that respondents who believed the coin toss was private would have answered more of the RRT items correctly, but that was not the

[5] The question wording was slightly different in the web mode: "During the survey, were you convinced that no one else would know whether the coin toss was heads or tails?"

[6] We omitted "Don't Know" responses from this analysis.

case. Respondents who answered Yes (believed it was private) answered 93% of the RRT items correctly (95% CI [0.875, 0.984]) in comparison with 97% of the respondents who answered No (95% CI [0.942, 0.990]). Respondents who believed the coin toss was private were not significantly more accurate on the RRT items, $t(58) = 1.36, p = 0.91$.

Two main themes emerged when respondents were asked to share any thoughts they had about the survey: (1) Respondents did not understand why we asked questions using the RRT, and (2) they enjoyed the process of flipping the coin to decide which question to answer. Ten respondents described the survey as "fun" or "interesting." One respondent explained, "I found it to be an interesting format. I routinely take online polls and surveys for a couple organizations. Yours seemed to involve me more, and require me to pay closer attention than those I'm used to." Another said, "I actually thought it was a pretty cool survey. I liked the virtual coin thing." See Chapter 11 on gamification for more information on this topic.

DISCUSSION

Our findings suggest two main conclusions: (1) Some respondents did not understand how to follow the procedure, and (2) those who did understand may have purposely *not* followed it at times. With a sample size this small and selective, these data are purely exploratory, but they do illustrate concerns with the RRT (e.g., Holbrook & Krosnick, 2010) that have been expressed previously. Further studies are needed to determine whether our statistically significant findings are in fact *practically* significant—i.e., do the findings extend to real-life applications?

Doing this research in Second Life allowed us to overcome real-life research barriers and take a closer look at how people responded to RRT items by using a method that is not possible in real life. However, the Second Life setting has some limitations: (1) Responding in Second Life may not be generalizable to responding in real life, (2) respondents may have doubted the coin toss was private or random, and (3) Second Life respondents may differ from the greater population in ways that affect RRT responding. Nonetheless, we consider the limitations to be rather minor for the following reasons.

We cannot be certain that the experience of participating in a survey and answering RRT items in Second Life is equivalent to doing the same tasks in real life. The Second Life setting may alter respondents' experiences. However, the first limitation, that responding in Second Life may not be generalizable to responding in real life, does not detract from our findings because respondents who demonstrated that they did not understand the procedure in Second Life are unlikely to understand it in real life. The research setting is more likely to have affected how respondents answered the social desirability questions. The finding suggesting that respondents purposely responded incorrectly in Second Life suggests that we should remain skeptical of the RRT in *any* setting, including real life. Even though we cannot be certain exactly how the Second Life setting changes the survey-taking experience, incorrect responding in Second Life is disconcerting.

The second limitation is that, as indicated by responses to the debriefing question, many respondents questioned whether the outcome of the coin toss was private. We do not believe this finding affects our conclusion that researchers should remain skeptical of the RRT. Respondents who believed the outcome of the coin toss was not completely private should have answered undesirable response items correctly because if they truly believed we knew the outcome of the coin toss, then it follows that we would know they were answering the innocuous question.

The third limitation is the possibility that the sample of Second Life respondents was not representative of the general population. Even if they differed, for instance on level of education (which is likely related to understanding of the RRT), the finding that they did not follow the RRT procedure is still troubling. Although this sample was drawn from among Second Life residents, the respondents are also in the universe from which samples are taken for real-life surveys. The sample in Second Life does not need to be representative before we should be wary of these sorts of findings.

Looking beyond this particular study, the same types of limitations and unanswered questions apply to other studies that use Second Life as a survey lab. It is unknown how Second Life respondents differ from real-life respondents, and whether they answer questions differently in Second Life. More specifically, do they take on the life of their avatar and answer as though they were someone else? Is avatar-to-avatar communication different from person-to-person communication?

Despite its limitations, Second Life is a valuable setting for experimental research. It allows researchers to control and manipulate the environment in ways that are not possible in real life. Even though this setting has limitations and the generalizability of findings may be uncertain, we have nothing to lose by doing exploratory studies in Second Life, so long as the results and limitations are considered in tandem.

REFERENCES

American Association for Public Opinion Research Cell Phone Task Force (2010). *New considerations for survey researchers when planning and conducting RDD telephone surveys in the U.S. with respondents reached via cell phone numbers.* Retrieved from http://www.aapor.org/AM/Template.cfm?Section=Cell_Phone_Task_Force_Report&Template=/CM/ContentDisplay.cfm&ContentID=3189.

AU, W. (2011). Second Life grows slightly in Q3 2011: Now 1.05M monthly users (good), 475K monetized users (better). *New World Notes.* Retrieved from http://nwn.blogs.com/nwn/2011/10/second-life-gains-monetized-users-in-q3-2011.html.

AU, W. (2012). Second Life stays on Nielsen's latest top 10 played PC games & increases total share, engagement. *New World Notes.* Retrieved from http://nwn.blogs.com/nwn/2012/07/second-life-top-pc-game-april-2012.html.

BAINBRIDGE, W. S. (2007). The scientific research potential of virtual worlds. *Science,* *317*(5837), 472–476.

CASTRONOVA, E., & FALK, M. (2008). *Virtual worlds as petri dishes for the social and behavioral sciences.* RatSWD Working Paper No. 47. Retrieved from http://ssrn.com/abstract=1445340 or http://dx.doi.org/10.2139/ssrn.1445340

COUTTS, E., & JANN, B. (2011). Sensitive questions in online surveys: Experimental results for the randomized response technique (RRT) and the unmatched count technique (UCT). *Sociological Methods & Research,* *40*(1), 169–193.

DAVIS, D. W., & SILVER, B. D. (2003). Stereotype threat and race of interviewer effects in a survey on political knowledge. *American Journal of Political Science,* *4*(1), 33–45.

DEAN, E., COOK, S., MURPHY, J., & KEATING, M. (2012). The effectiveness of survey recruitment methods in Second Life. *Social Science Computer Review,* *30*(3), 324–338.

DONOVAN, J. J., DWIGHT, S. A., & HURTZ, G. M. (2003). An assessment of the prevalence, severity, and verifiability of entry-level applicant faking using the randomized response technique. *Human Performance,* *16*(1), 81–106.

HARRISON, G. W., HARUVY, E., & RUTSTRÖM, E. E. (2011). Remarks on virtual world and virtual reality experiments. *Southern Economic Journal,* *78*(1), 87–94.

HOLBROOK, A. L., & KROSNICK, J. A. (2010). Measuring voter turnout by using the randomized response technique: Evidence calling into question the method's validity. *Public Opinion Quarterly,* *74*, 328–43.

KANE, E. W., & MACAULAY, L. J. (1993). Interviewer gender and gender attitudes. *Public Opinion Quarterly,* *57*(1), 1–28.

KRYSAN, M., & COUPER, M. (2003). Race in the live and the virtual interview: Racial deference, social desirability, and activation effects in attitude surveys. *Social Psychology Quarterly*, *66*(4), 364–383.

LARA, D., STRICKLER, J., DIAZ OLAVARRIETA, C., & ELLERTSON, C. (2004). Measuring induced abortion in Mexico: A comparison of four methodologies. *Sociological Methods Research*, *32*(4), 529–558.

LENSVELT-MULDERS, G. J. L. M., HOX, J. J., VAN DER HEIJDEN, P. G. M., & MAAS, C. J. M. (2005). Meta-analysis of randomized response research: Thirty-five years of validation. *Sociological Methods Research*, *33*(3), 319–348.

MURPHY, J., DEAN, E., COOK, S., & KEATING, M. (2010, December). The effect of interviewer image in a virtual-world survey. RTI Press Publication No. RR-0014-1012. Research Triangle Park, NC: RTI Press.

SHEPHERD, T. (2012). *Second Life Lindex daily market data repository*. Retrieved from http://gridsurvey.com/lindex.php.

TOURANGEAU, R., & YAN, T. (2007). Sensitive questions in surveys. *Psychological Bulletin*, *133*(5), 859–883.

TOURANGEAU, R., COUPER, M. P., & STEIGER, D. M. (2003). Humanizing self-administered surveys: Experiments on social presence in Web and IVR surveys. *Computers in Human Behavior*, *19*, 1–24.

WIMBUSH, J. C., & DALTON, D. R. (1997). Base rate for employee theft: Convergence of multiple methods. *Journal of Applied Psychology*, *82*(5), 756–763.

ZDEP, S. M., & RHODES, I. N. (1977). Making randomized response technique work. *Public Opinion Quarterly*, *40*, 531–7.

ZDEP, S. M., RHODES, I. N., SCHWARZ, R. M., & KILKENNY, M. J. (1979). The validity of the randomized response technique. *Public Opinion Quarterly*, *43*, 544–9.

CHAPTER 7

Decisions, Observations, and Considerations for Developing a Mobile Survey App and Panel

David Roe, Yuying Zhang, and Michael Keating,
RTI International

Survey research has had to adapt to advances in technology and changes in society. From the development of telephone interviewing as a supplement, and later an alternative, to mail and face-to-face interviewing, to web surveys, to the inclusion of cell phones in random digit dial (RDD) surveys, the field of survey research has worked to determine the best uses of new technology. Now, researchers find themselves in the midst of another wave of change that they must work to understand fully and use: mobile data collection. The landscape of survey research is changing drastically as a result of advances in mobile technologies and increased accessibility.

In addition to improvements in coverage, speed, functionality, and computing power, smartphone applications (apps) offer a robust set of features to researchers. Understanding the art of capturing data via

Social Media, Sociality, and Survey Research, First Edition.
Edited by Craig A. Hill, Elizabeth Dean, and Joe Murphy.

mobile surveys, specifically via applications, will be an important task that the survey research industry must address.

In the past 3 years, survey research has addressed the evolution of mobile devices and the impact of smartphones. However, most early studies focused on how to handle users accessing existing surveys on mobile devices, not on how to deliver the surveys, that is, via a mobile web browser or some type of application. Initial research focused on mode effects, layout, usability, question wording, processing, form factors, etc. (see Buskirk & Andrus, 2012; Wells et al., 2012; Peytchev & Hill, 2010; Couper, 2010; Callegaro, 2010; Dawson, 2010; Koch, 2010). Although extremely important to the overall understanding of the impact of mobile devices, these studies dealt more with the item-level methodological considerations of this new data capture mode without stepping back to determine how to optimize mobile survey operations. By mobile survey operations, we mean the method by which we distribute surveys, collect data, and manage respondent participation and progress (in either a single survey or a panel). Today, we have far more insight into the habits of users, which makes solving the riddle of optimization a bit, but not much, easier. At the very least, as the knowledge of user habits grows, survey researchers can begin to develop research questions that focus on mobile survey operations. For the research team in this chapter, the questions focus on the use of mobile applications (apps) for data capture.

In a recent report, the Online Publisher's Association (OPA) stated that 96% of smartphone content consumers had downloaded apps in the past year (2012). More and more mobile apps are being developed every day to target this growing market, in both commercial and noncommercial environments. Even traditional software applications are forced to have a mobile presence to avoid becoming obsolete. This chapter provides background and thoughts on the impact and evolution of data collection technology; presents factors that should be considered when developing an application and managing a panel of users; examines preliminary data from the public release of a data collection app; and suggests topics and directions for future research.

IMPACT OF THE EVOLUTION OF TECHNOLOGY ON DATA COLLECTION

Over the past 75+ years, the survey research industry has seen major changes in the technology available to optimize data collection quality and efficiency. This technology evolution, as applied to survey research, can be compartmentalized into three key devices or platforms: telephone data collection, web or online data collection, and mobile data collection. In each of these areas, the survey research industry has had to adapt to and face methodological challenges rapidly.

Telephone Interviewing

Consider first the growth of telephone interviewing as a supplement, and later an alternative, to mail and face-to-face interviewing. Originally used as a method of support for other interviewing techniques, the telephone was employed in combination with mail or face-to-face interviews, usually for purposes other than gathering information from respondents (Frey, 1983). Eventually, telephone interviewing evolved and became a stand-alone method and a more practical alternative for interviewing representative samples of the general population. Furthermore, the potential cost savings telephone interviewing offered when compared with face-to face interviewing also made it a viable alternative (Bardes & Oldendick, 2000). That said, its evolution took time, with coverage expanding from 35% to 95% of households between 1936 and 1996 (Bardes & Oldendick, 2000).

Web Interviewing

That 60-year period looks slow compared with the rapid growth of other modes now used in data collection. First, consider the rapid evolution of the Internet itself. In a fraction of the time it took for survey research techniques using the telephone to evolve, the World Wide Web has reached relatively high levels of penetration throughout the world. Internet World Stats (2012) estimated that Internet penetration (i.e., the number of Internet users divided by the population and expressed as a percentage) had reached 34.3% worldwide by the end of the second quarter of 2012. Although that number may seem small now, one must also consider the growth behind it.

The 2012 penetration statistic represents over a 500% rate of growth from 2000 to the second quarter of 2012. And the growth is even higher in developing countries, often viewed as the source of most "gaps" in the penetration of new technologies. For example, Africa's Internet penetration has grown by over 3,600% in the aforementioned 12-year period, while penetration in the Middle East and Latin America has grown by

2,600 and 1,300%, respectively. In the United States, the audience of Internet users was projected to grow to approximately 239 million by the end of 2012, representing 75.6% of the total population (Vaughan, 2012). And, as coverage has spread, the evolution of Internet research has followed suit, and the adoption of online surveys has spread faster than any other similar innovation (Couper, 2010).

Cell Phones

Several growth spurts in survey research have taken place as a result of developments in the technologies that link the researcher to the respondent. Although the advent of Internet surveys created issues of noncoverage bias and other errors in web surveys themselves (see the American Association of Public Opinion Research [AAPOR] 2010 Online Panel Task Force Report for a wealth of examples), the increasing number of cell phone owners and the resulting proportion of cell-mostly and cell-only households and individuals in the United States struck quite a blow to RDD telephone surveys. And it happened quickly.

In 2010, the National Center for Health Statistics (NCHS) estimated that 25% of U.S. households and 23% of adults had no landline service and were cell only by the end of 2009, which is a significant increase in both statistics since 2006 (Christian et al., 2010). More time passed, and the growth continued. In a recent publication, NCHS estimated that more than one third of American homes (35.8%) had only wireless telephones (Blumberg & Luke, 2012). Although these statistics help us to understand why bias in landline samples is increasing, delving deeper into the data provides even greater cause for concern. Demographic differences or gaps continue to widen across a wide variety of demographics, but four key groups stand out: adults aged 25–34, adults living with unrelated adult roommates, adults renting their home, and adults living in poverty (Blumberg & Luke, 2012). In addition to these key demographics, cell-only adults were significantly different from adults living in landline households when it came to substantive measures, such as alcohol consumption, tobacco use, certain chronic diseases, and access to health care (Blumberg & Luke, 2012). These differences are critically important. Researchers who exclude cell-phone-only sample members from telephone surveys could be eliminating members of the population that are far different from their landline counterparts, resulting in less than representative results.

From the few examples cited here, it is easy to see how the cell phone has affected survey research efforts. It has impacted, and in some ways threatened, how survey researchers capture representative data. And, in the spirit of evolution and change, we know now that a particular subgroup of cell phones, smartphones, represents the majority of device types in the United States (Nielsen, 2012) and is causing researchers to rethink once again how to adapt to and leverage technological innovation.

Smartphones

In examining this increase in cell phone ownership, we must also determine how emerging technologies and device ownership can best be used for the good of survey research. Perhaps the greatest potential benefit to survey researchers interested in harnessing the power of emerging technologies is the rapid growth of smartphone ownership. A smartphone is defined as a mobile phone built on a mobile operating system, with more advanced computing capability and connectivity than a feature phone, which has additional functions over and above a basic mobile phone capable only of voice calling and text messaging (Phonescoop, 2012).

The growth in the speed, computing power, and capabilities of these devices is staggering. In 1965, Gordon Moore, one of the founders of Intel, stated what became known as Moore's Law: Computer power doubles every 18 months. Need proof of this concept? Consider this excerpt on the power of cell phones from a recent book by physicist Michio Kaku (2011, page 21):

> *Today, your cell phone has more computer power than all of NASA back in 1969 when it sent two astronauts to the moon. Video games, which consume enormous amounts of computer power to simulate 3-D situations, use more computer power than mainframe computers of the previous decade. The Sony PlayStation of today, which costs $300, has the power of a military supercomputer of 1997, which cost millions of dollars.*

As these devices have become more powerful, they have become more popular. The number of cell phone users now using smartphones is also on the rise. In 2011, 38% of mobile phone users in the United States owned a smartphone (Buskirk & Andrus, 2012). By the end of the third quarter of 2012, a majority of mobile phone users (56%) owned smartphones, up from 49% in the first quarter of 2012 (Nielsen, 2012).

Over the past few years, as smartphone coverage in the United States has increased, and associated technologies have become increasingly more powerful, many predicted that smartphones would be able to perform many of the functions of current desktop computers and laptops, leading more people to access the Internet on mobile devices than on laptop or desktop computers. These predictions seem to be coming true. As of April 2012, 55% of adult mobile phone owners reported using the Internet on their mobile phones, nearly double the percentage in 2009. Furthermore, 31% of these current cell Internet users report that they do most of their online browsing using their cell phone, instead of using another device like a desktop or laptop computer; in other words, 17% of all adult cell owners are "cell-mostly Internet users" (Smith, 2012).

The growth statistics are exciting, but survey researchers must be prepared to deal with differences in adoption rates when considering how data might be captured on smartphones moving forward. For example, the Pew Research Center notes that smartphones are particularly popular with young adults and those living in relatively higher income households (Pew, 2012). Although this difference in demographics might give survey researchers pause when it comes to attempting to collect representative data via a smartphone, these differences may disappear in the near future, especially in the United States. In a 2012 article, Henry Blodget of *Business Insider* suggests that because the adoption of smartphones has taken place so rapidly, the point of full penetration may be only a few years away. Blodget notes that with this increasing penetration, the demographic differences witnessed in 2012 will continue to erode, and the as-yet unpenetrated segment of the U.S. smartphone market—the late majority adopters who will likely buy smartphones for the first time over the next few years—will be older and less well off than those who already own smartphones.

Smartphones are also changing users' habits, behaviors, and expectations. An article on wireless technologies in a recent issue of *Time* magazine noted that, "it is hard to think of any tool, any instrument, any object in history with which so many developed so close a relationship so quickly as we have with our phones" (Gibbs, 2012, paragraph 1). The article goes on to note that for many, a smartphone is the first phone they have ever had. And access to this device has changed the way we behave. As Figure 7.1 shows, "[a] tool our parents could not have imagined

Our Smartphone Habits

Smartphone ownership reached a tipping point in 2012, where more Americans are buying them to take advantage of applications and other sophisticated web-enabled features compared to traditional cell phones. Pew Research Center's Pew Internet & American Life Project has been tracking trends in cell phone ownership, use, and its role in the digital divide for the past eight years.

Percent of U.S. adults who own a smartphone

35%   **45**%

May. 2011   Sept.. 2012

Nearly half of all American adults (45%) and two-thirds of all young adults now own a smartphone. In 2012, our data showed that smartphone owners are now more prevalent within the overall population than owners of more basic mobile phones.

Smartphones are particularly popular with young adults and those living in relatively higher income households.

66%

of those ages 18-29 own smartphones...

 $75,000

...68% of those living in households earning $75,000 own Smartphones.

A majority of adult cell owners (55%) now go online using their phones, and 17% of cell phone owners do most of their online browsing on their phone, rather than a computer or other device. Most do so for convenience, but for some their phone is their only option for online access.

As of May 2011, texting and photo-taking are the most common activities for all mobile users.

41.5 messages exchanged by median adults each day.

Texting remains much more popular among young adults.

109.5 messages exchanged each day by young adults aged 18-24.

Call me, maybe?

One half of cell owners say they prefer someone call them to contact them; 1/3 said they prefer a text message

---

Nearly half of smartphone owners said they have used their phone in the past 30 days to:

Look up something to settle an argument.

Decide whether to visit a business, such as a restaurant.

Solve an unexpected problem.

PEW INTERNET & AMERICAN LIFE PROJECT

FIGURE 7.1 Our smartphone habits.

has become a lifeline we can't do without. Not for a day—in most cases not even for an hour. It is a form of sustenance, that constant feed of news and notes and nonsense, to the point that twice as many people would pick their phone over their lunch if forced to choose" (Gibbs, 2012, paragraph 4).

Just as smartphone users' behaviors are changing, so too are users' expectations of content delivery and interaction. Survey researchers wishing to leverage this medium for data collection must recognize that over the past few years, the way consumers access content and information as well as stay informed and entertained has shifted. Consumers now expect that they will have the world at their fingertips anytime, anywhere, which has led to publishers working feverishly to optimize their mobile sites and create apps that meet these expectations (OPA, 2012). In this context, survey research on smartphones must be viewed as almost any other application or opportunity for interaction would be viewed. The smartphone has become such a personal device, and has delivered information in such a paradigm-shifting way, that survey researchers can no longer focus only on developing and maintaining research methods for stand-alone data capture techniques. Researchers who wish to collect data via emerging smartphone technology must design protocols that effectively blend with all of the other commercial applications and media that smartphone users are accustomed to, including look, feel, speed, and maintenance of interest and engagement.

BUILDING AN APP

Many companies run survey websites and panels to conduct market research on their product offerings. However, application-based offerings created by what most would consider the traditional survey research industry are less prevalent. Contract research organizations, federal agencies, academic research centers, and others that rely on scientifically sound, robust, and generalizable results to provide clients with information that may shape policy, funding decisions, and other important considerations are beginning to turn their attention to the development of survey apps.

To this end, our research team decided to embark on the creation of an app, SurveyPulse™, by RTI International (SurveyPulse). SurveyPulse is a mobile application designed to deliver surveys to users across multiple platforms, operating systems, and devices including tablets and to collect data in real time. In this section we share the goals, key decision points, findings, and lessons learned associated with the process of getting the app off the ground, from design choice, to operational decision making, to working with internal organizations such as legal departments and institutional review boards (IRBs) to make the app possible.

Goals

The research team that developed SurveyPulse had several goals for this endeavor. We sought to develop and study a panel of survey users who completed surveys for small rewards to identify issues associated with this mode while experimenting and investigating a number of methodological questions. We hoped to learn about establishing a panel that has as many users as possible, and eventually uses a sampling approach more rigorous than open adoption or simple opting in. Several additional experiments and investigations are also planned. Areas of exploration include:

- Recruitment—If we leave panel enrollment to chance, by simply placing the app in stores for users to download, what does our user base look like? Is it demographically skewed to certain types of users? Can the use of probability-based sampling—specifically address-based sampling (ABS)—lead to a more representative user base?

- Attrition and engagement—At what point do respondents lose interest in participation? What can be done to enhance the user experience to maintain the size of the panel or limit attrition?

- Incentives—At what point does return on the use of respondent incentives cease? Which methods, such as paying per single surveys versus groups of surveys, result in higher response rates and better quality data?

- Questionnaires—How do researchers ask the questions they or their clients are interested in while also asking the questions respondents want to answer, in order to keep them engaged and interested?

- Survey length—How long is too long on a mobile device? Are certain questionnaires bound to fail on the mobile platform based on size, length, or perceived burden?

- Survey frequency—Just as survey length can affect burden and attrition, the frequency with which we ask respondents to provide data may also prove problematic. In addition, it will be important to understand how frequency and survey length interplay with one another. For example, are longer surveys better broken up over time, or should they be administered with a buffer around them, keeping respondents from feeling like they have been asked to do too much, too quickly?

- Respondent communication—What is the best way for researchers to communicate with panel members? Can we keep users engaged through in-app communication, or are additional outside interventions such as mailings, phone calls, or text messages critical to engagement?

Evolution of an Idea: Critical Decision Points

As the research team quickly found out, the evolution of our initial ideas and goals faced critical decision points along the way. Do we build our own app or buy into an existing panel, leaving us less leeway to experiment with mobile operations? Do we build only for one operating system to start, or do we try to incorporate more users via at least two systems? How do we recruit users and keep them engaged, and how do we help to educate our own internal sources of support to ensure that what we create will best represent the organization?

Decision Point 1: Build It or Buy It? Survey researchers interested in diving into these new modes usually come to a key decision point quickly. One option for researchers is to attach surveys to existing third-party applications offered by companies that allow them to work within an established infrastructure. Alternatively, researchers can decide to invest their own resources into developing and customizing their own data capture application. In addition to time and money, other key aspects of data collection have to be considered. Survey researchers face a twofold challenge when designing new approaches to survey data collection. First, to ensure that quality data are captured, survey researchers must follow sound methods rooted in theory and best practices. Furthermore, survey researchers are ethically and legally obligated to protect the rights and privacy of all participants, including both sample members and respondents.

Keeping these factors in mind, researchers need to decide early in the development process whether to build the app or buy it. A key question is as follows: Does an app already exist that meets your standards and allows you to control key pieces of the data capture process, or are you willing and able to invest the resources a custom application requires? We considered this question and began developing Survey-Pulse by researching and evaluating a number of applications through both materials review and interviews and discussions with software vendors. These evaluations included:

- Supported platforms and operating systems
- Costs, hosting, and licensing
- Company history, reputation, and user reviews
- Maintenance and support for developers and users
- Ability to customize user interfaces
- Usability

The following section discusses some critical points and additional details that researchers should consider when they face this decision.

Does the Application Work Across Multiple Smartphone Platforms? App development costs money. Researchers who choose to develop their own apps are creating a tool from the ground up in an ever-changing environment. It is easy to fall into a trap of developing an app for one platform over another, and some research panels using apps are currently only operating on a single platform. However, data on the diversity of smartphone platforms suggest that operating on a single platform contributes to bias issues associated with undercoverage (defined by Groves et al., 2004, as the occurrence of a population of interest differing from the sampling frame).

The statistics on operating system use suggest that choosing one system over another can be dangerous. At the end of the third quarter of 2012, Nielsen Mobile Insights released the following information on market share. Figure 7.2 shows that just over half (52%) of the U.S. smartphone market belonged to the Android operating system, while Apple's iOS followed with 35% of the market. Blackberry (7%), Windows Phone (2%), and others (5%) rounded out the market. These

Top U. S. Smartphone operating systems by market share
Q3 2012, Nielsen mobile insights

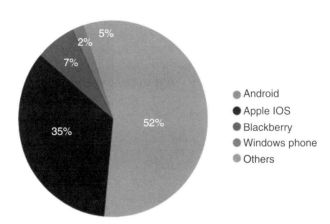

Read as: During Q3 2012, 52% of smartphone owners had
a handset that runs on the android operating system

Source: Nielsen   nielsen

FIGURE 7.2   Top U.S. smartphone operating systems by market share.

numbers suggest that leaving out over a third of the market by developing
only for Android or half of the market by developing only for Apple is
a mistake.

Before these numbers were public, a Pew study had noted demographic differences by platform use, finding higher rates of Android
adoption among African Americans and young adults and high ownership rates for iPhones among those with higher educations and household incomes (Smith, 2012). This point is critical, as excluding users on
a certain platform such as Android or iPhone would limit the ability to
capture information from a diverse population of respondents.

How Is Data Security Being Addressed? As stated earlier,
survey researchers have a responsibility to protect respondents even
while exploring innovation. If researchers choose an existing vendor,
that vendor must be transparent with data security. IRBs and other
entities that protect human subjects are very concerned about protecting
personally identifiable information. This issue could be a roadblock for
researchers who want to use third-party developers' services or develop

their applications on their own. This is especially true if no option exists to transmit and store data safely on the researchers' servers or networks.

How Much Training Is Required to Learn and Use Existing Applications?
Will purchasing an app create the need for developers and survey managers to learn a new system? If so, the time it takes to learn how to program new surveys (if an option), upload content, monitor data collection metrics, translate and manipulate data, and so on might be cost prohibitive when compared with in-house development of native code, in which existing expertise could be applied from the start. In addition to the labor involved for developers and managers, a lack of understanding of the inner workings of a certain app could damage the researcher–respondent relationship. Often, survey researchers provide substantive and technical support to their sample members. Doing so is critical to ensuring that respondents can overcome technical difficulties they may experience while attempting a self-administered interview. Respondents may also misunderstand, misinterpret, or be confused about the meaning of a given survey question, and they would not have a trained professional to turn to for clarification as they would during a telephone interview. If researchers are the only source of help for respondents using an app from another company, an inability to understand quickly the software being used could hurt the credibility of the research study.

How Quickly Will Updates and Fixes Become Available?
Just as being able to address questions or concerns from respondents is critical, providing the best possible app and experience for respondents is equally important. Part of respondent engagement relies on a positive experience. If an app requires updates and fixes so respondents can use it, researchers have to consider the impact on respondents' perception of the app's reliability and usability. The key question is as follows: If you recognize usability issues, can you fix them in a timely manner that works best for your respondents and your data quality?

Can the App Be Integrated with Existing Survey Systems?
Can the app, either built or bought, be integrated with other modes of data capture that might be needed for multimode studies? This question is crucial at the organizational level. The researcher has to know whether the organization would ever need this app to be one piece in a larger puzzle of systems management or data collection management tools.

Does the App Provide Branding? Often, survey research protocols consist of multiple steps that all tie to routing a sample member
to the final product, the data capture tool. Advance letters, envelopes,
postcards, and flyers all carry a brand to establish and maintain the
credibility of a research study while preparing respondents for future
communication. In our investigation, we found that not all apps available for purchase allow researchers to brand their app. If it is important
for researchers (or their clients) to present data capture tools with the
logos and symbols associated with the organization or client to maintain credibility and salience for respondents, this aspect of evaluating a
solution becomes critical. In the end, our team decided that it was best
to begin developing our own application.

Decision Point 2: What Do We Build? Even as this critical build
decision was made, we had to make an additional choice—what kind
of app do we build? We already knew we were going to develop an
actual app, as opposed to trying to work with web surveys optimized for
mobile devices. Because every mobile device comes with a built-in web
browser, one might think that developing a website optimized for mobile
users would be the easiest way to reach a large audience. However,
people are used to apps, not mobile browsers, and researchers must be
sensitive not only to coverage and exposure but also to expectations.
Furthermore, the best way to build and maintain such a website has not
been established. Some tools are now available to make a website look
and perform well on mobile devices, but no clear universal solution
exists.

A hybrid application combines elements of both native apps and web
applications. It is typically a web application wrapped in a native web
view container with some limited native enhanced features. A hybrid
app is a good compromise when quicker development across multiple
platforms with native features is needed. A hybrid application behaves
and appears like a native mobile application even though internally it is
a mobile web application.

At the highest level is the generated native application. Mobile optimized websites and hybrid apps have their weaknesses, from reduced to
no exposure in public venues like the Apple App Store or Google Play,
to limits on smartphone features such as camera, video, communication, and data storage. Although native mobile apps are perceived to be
superior to mobile websites in many situations, they pose an additional

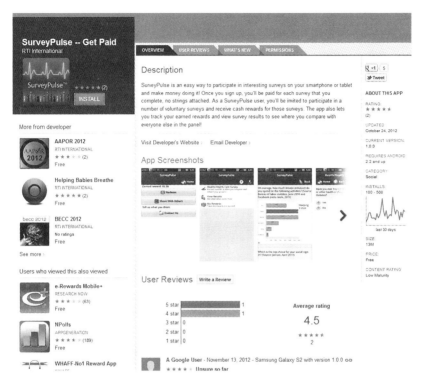

FIGURE 7.3   SurveyPulse download page in Google Play.

challenge to researchers and developers. Developing apps for Android devices is completely different from developing apps for Apple iOS devices. The amount of effort required to develop and maintain an app for both Android and iOS, ensuring that we are doing all we can to avoid the coverage issues discussed earlier, is sometimes double that of mobile optimized websites. However, the good news is that third-party tools are available that allow developers to write in one language and then generate native apps for multiple platforms. Using this strategy in the pilot stage, we created SurveyPulse, a single application that works on both the Android and iOS platforms, with no differences in appearance or function (see Figures 7.3 and 7.4).

Decision Point 3: Who Do We Recruit and How?   Recruitment had two phases: The first phase was open adoption, whereby users joined the panel on their own via app store searches, referrals from friends and

FIGURE 7.4   SurveyPulse download page in iTunes.

family, and so on. It is no way meant to be representative. Phase 2, implemented in early 2013, used a probability-based approach, address-based sampling (ABS), to explore the potential of gaining a more representative pool of users, increasing coverage and reducing bias.

ABS frames are derived from commercially available versions of the U.S. Postal Service (USPS) computerized delivery sequence (CDS) file and what is known as the No-Stat file, a file of over 8 million primarily rural mailing addresses that supplement the CDS file with both active and vacant addresses that are excluded from the CDS file. The union of the CDS and No-Stat files accounts for all postal delivery points serviced by the USPS, giving ABS frames near-complete coverage of the household population. In addition, an ABS design has several statistical advantages that could help establish a representative panel as smartphone penetration continues to increase. Because addresses are selected from a single frame, sample designs are more efficient, enabling more precise survey estimates. Thus, fewer respondents are needed to achieve the same level of precision in a traditional RDD design. This single frame also simplifies weighting thanks to the elimination of multiplicities across frames (e.g., persons with both landlines and cell phones). Because the locations of sampled households are known, ABS designs can also implement more efficient oversampling of minority populations to target the desired minority percentage in each market. Finally, the ABS approach provides the locations of both responding and nonresponding households, so better nonresponse adjustments can be implemented.

Admittedly, our expectations were very low as the first recruitment phase began. With no advertising or direct contact with potential respondents, we expected our app to be lost in a sea of apps. As mobile connectivity has increased, so has the opportunity for developers to try their hand at developing applications for mobile devices. Apple offers iOS users over a half-million apps, and in early 2012, they proudly announced that the app store topped the 25 billion download mark (Apple, 2012). Google's Android market is not far behind. In early 2012, an estimated 400,000 apps were available through Google Play, the newly titled Android market (Calloway, 2012).

Along with this boom in the creation and availability of apps comes the potential for saturation and fatigue. Apps can easily get buried or overlooked among the hundreds of thousands of other apps: Making a top-25 list is very difficult. If an app is not titled or described in the best possible way, it could get lost amidst competitors whose keywords, descriptions, or icons might help them get noticed faster. Indeed, there are now companies devoted to helping developers get their apps noticed.

More importantly, users can become easily overwhelmed by the choices of apps they have for any given task or activity. By using more traditional survey sampling methods to get the word out and recruit, we hope to circumvent the challenges of being picked up in an app store, but for now, we are happy to put the app in the stores and begin to take advantage of this complex opportunity.

Decision Point 4: How Do We Keep Users Engaged? In preparing to develop and pilot SurveyPulse, our research also revealed that users can become easily overwhelmed by the variety of apps with the same ostensible purpose. Users are hitting a wall when it comes to the myriad options they have for apps, so much so that developers are now producing an app to help mobile users find apps (McHugh, 2012). Furthermore, even if users download an app, they may not use it regularly. In fact, the average consumer has about 65 apps and yet only uses 15 per week (Perez, 2011).

The lack of regular use or regular attention is troubling to researchers when they try to plan the development of an app for survey data collection. In maintaining a panel, survey researchers have to ensure that the app itself does not get lost among more popular apps like Instagram and Angry Birds. Simply paying panel members to participate is not necessarily effective. Innovations that provide benefits to the user, such as real-time results, user rankings, quickly available surveys, and more must be in place to enhance users' experience and keep them engaged. As shown in Figures 7.5–7.7, we developed SurveyPulse with some enhanced features, such as a home screen to show available surveys; reward tracking; and data sharing, in which we post aggregate results of past surveys, allowing users to track the rest of the panel and see where they fit among others.

Decision Point 5: Are We Ready? At this point, developers have made many decisions. The app has been programmed, some test surveys have been written, and testing its basic functionality has concluded: The app is ready to be launched. But what about all of the other organizational factors that a research team must consider? Using an app is novel not only in terms of data capture, but also it is new in organizational aspects, namely, legal considerations, public relations, and most importantly, respondent protection.

Legal Considerations. We learned quickly that the informed consent scripts we were accustomed to using many surveys would not be enough for SurveyPulse. In addition to consultation with our Office of Research Protection, meeting with corporate counsel was required. The resulting terms and conditions agreement far surpassed the length and content of typical informed consent, covering information on our organization, who can participate, rights and privacy, our purpose, data security, user safety, schedule, compensation, and the right to extend the study. Luckily, our legal team guided us through the process; so survey researchers should have guidance in understanding this new area before they proceed.

Public Relations. As a team, we thought we were in great shape. We had an app, a logo, and a snappy name. Our legal terms and conditions were looking good, and we were ready to upload to the stores and see what would happen. But before we did, we also needed to ensure that our efforts satisfied the expectations of our organization's public relations and communication standards. Was the organization being represented properly? Were we using colors, logos, and so on appropriately? Were our online accounts centralized for all developers? Did we have a streamlined communication process in place so that users

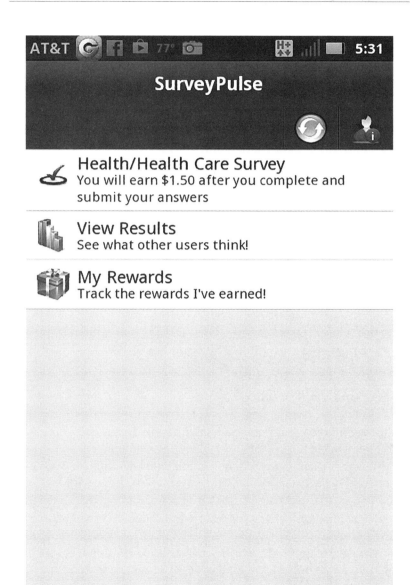

FIGURE 7.5   SurveyPulse home screen.

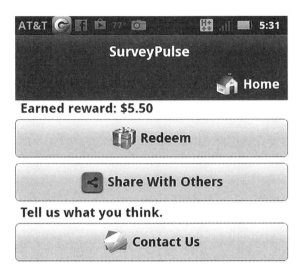

FIGURE 7.6 Reward tracking.

could reach out for help if needed? In our case, the answers were yes, but it was incredibly helpful to make sure that all of these bases were covered with the help of experts beyond survey research.

Respondent Protection. Just as this type of data capture was new for the survey researchers, it was also new for our IRB and Office of Research Protection. Essentially, the standard protocols for survey data collection did not completely address the requirements for efficient review and respondent protection in the mobile environment. Although most of the issues were resolved through conversations with our legal support team, the study team and IRB had to work together to develop an understanding and a plan for what was needed to provide adequate review and approval of study protocols. For example, consider one of the most attractive features of this type of app, the ability to release surveys quickly to select individuals to gain rapid insight on a current event. To acquire timely IRB approval for these types of surveys, a specific, streamlined review plan had to be in place. Only through discussion and heavy planning was this possible.

PRELIMINARY FINDINGS

In this section, we share some of the results we have experienced since the launch of the SurveyPulse app in October 2012. As a reminder, our

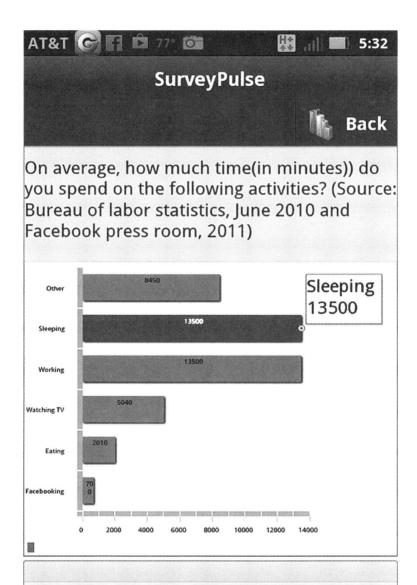

FIGURE 7.7   Example of result sharing.

results come from the first phase of recruitment, which was based on open enrollment. Furthermore, although not all of the research goals discussed earlier can be fully analyzed at this time, beginning to gain insight directly from users is also critically important. Although these results are not representative, they are interesting in the context of next steps in the evolution of mobile survey operations.

Recruitment

Because of the lack of outreach involved in simply adding the app to user markets, we had low expectations about the number of adopters we would receive. However, much to our surprise, to date, the SurveyPulse app has been downloaded just over 800 times. Once duplicate downloads are removed, the number is closer to 750. We anticipated duplicate downloads and registrations, and as a result, these cases are managed on the back end or server side, after examination of the data.

For the purposes of early analysis, users are described as panel members who not only downloaded the application and completed the registration process but also participated in the first full survey made available to them, a demonstration survey designed to walk them through using SurveyPulse. In total, 654 users completed the first survey and are currently considered members of the SurveyPulse panel. As discussed earlier, coverage issues can arise if preference is given to one operating system over another. Our numbers actually differ from the market share statistics. As shown in Figure 7.8, panel members using iOS devices make up just over 60% of membership, while the rest are Android users.

As expected, the demographics of the panel as a result of open adoption are not very representative, as illustrated in Figures 7.9–7.13. Almost three quarters of the current panel members are between the ages of 18 and 34, while only 2% are age 55 or older. Members in the majority are primarily white (67%), female (71%), have at least some college education (58%), and earn $35,000 a year or less (60%).

Respondent Communication

When they registered, users were asked to provide some information for the sake of respondent communication, invites, and updates. Respondents were asked to provide their e-mail address and cell phone number and whether they prefer to receive survey notifications (invites and

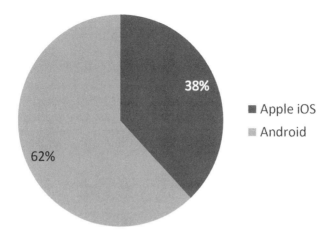

FIGURE 7.8   SurveyPulse users by operating system.

reminders) or not. Over half of the panel members (55%) provided an e-mail address for contact, while approximately 42% provided a cell phone number (Figure 7.14). For notification preferences, 72% of panel members indicated that the survey team could contact them regarding new surveys, while the remainder indicated that they would check for new surveys themselves (Figure 7.15). Of those who agreed to be

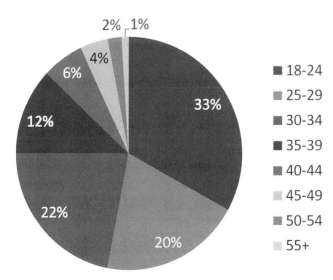

FIGURE 7.9   SurveyPulse users by age.

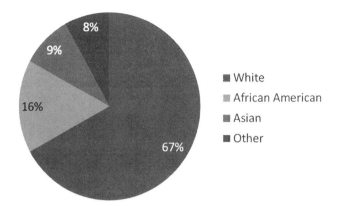

FIGURE 7.10 SurveyPulse users by race.

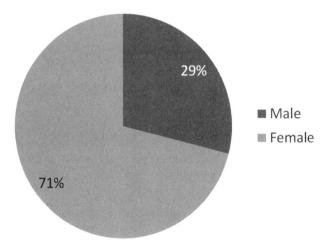

FIGURE 7.11 SurveyPulse users by gender.

contacted, 46% preferred e-mail contact, while 30% indicated that they would prefer a text message. The remaining 24% indicated that e-mails or texts would be acceptable (Figure 7.16). Future analysis will look across all panel members and surveys to ascertain whether one mode of communication was more successful than another.

Survey Topics

Our research also seeks to ask questions respondents want to answer, in order to keep them engaged and interested. Preliminary results show

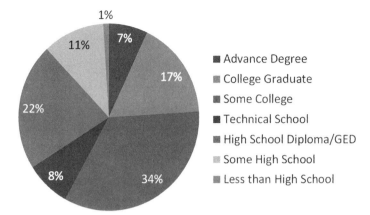

FIGURE 7.12   SurveyPulse users by highest level of education completed.

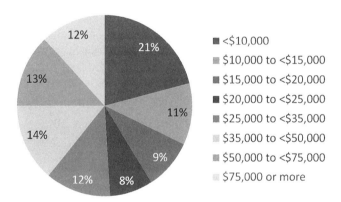

FIGURE 7.13   SurveyPulse users by income.

that, generally speaking, surveys on current events have better completion rates than surveys on more general topics. For example, although
a survey on credit card use and debt had a 25% completion rate, a
survey conducted the Monday after the tragic events in Newtown, CT
(concerning the same subject), on Friday, December 14, 2012, had a
97% completion rate. And, although future investigations must explore
other confounding factors, such as incentive differences between all surveys, among other factors, it should be noted that the credit card survey
offered respondents a $2 incentive, while the Newtown survey offered
only 75 cents.

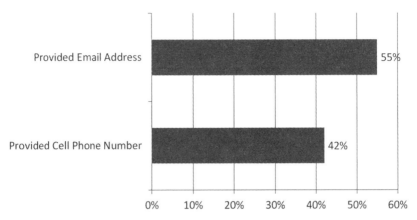

FIGURE 7.14   SurveyPulse contact information provided by users.

Other current event surveys, such as one conducted on college football bowl season in mid-December 2012 (37%) and another conducted
on the fiscal cliff in late December (30%), brought in higher completion rates than other surveys focused on more general topics and not
current events. These results are not representative, and the research
team fully recognizes their preliminary nature, but as mobile survey
operations evolve, survey topics must be included in the conversation when considering how to maintain a panel and keep members
engaged.

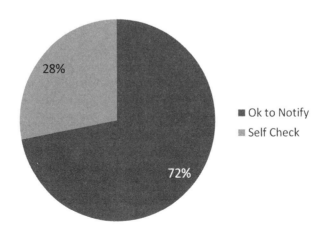

FIGURE 7.15   Notification preferences of SurveyPulse users.

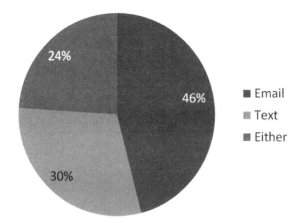

FIGURE 7.16 Notification method preferences of SurveyPulse users.

Respondent Impressions on Incentives, Survey Length, and Frequency

As the work on SurveyPulse progresses, panel members will be given an opportunity to comment on their experiences. This type of assessment is critical to the learning process; the qualitative assessment about incentives, survey length, and survey frequency will help us understand how to optimize mobile survey operations. To date, only a very small number of members have provided such information. That said, early results suggest a preference toward multiple surveys in a given week and preferences toward long surveys with higher incentives. Furthermore, open-text responses have shed light on the need to consider and investigate members' desire for multiple cash-out options (such as PayPal) and linking results and experiences to social media.

NEXT STEPS

Continued exploration and innovation are needed for mobile data collection. Clearly, more work must be done, especially concerning probability sampling and the penetration of smartphones in the broader population, which would suggest a greater likelihood of capturing representative data. In addition, researchers will be well served to make sure that the user is truly considered. Unlike computer-assisted telephone interviewing or paper-and-pencil interviewing, mobile data collection is not a stand-alone mode. SurveyPulse is competing with other applications, and our respondents' expectations will be shaped by their experiences with other apps.

This early look at reactions, responses, and behaviors related to using SurveyPulse are exciting. However, we must further explore and understand the potential pitfalls associated with data collection via an app. First and foremost, although open adopters came in far greater numbers than we anticipated, phase 2 of recruitment, using an ABS approach, is critical to understanding in greater detail how a representative panel of mobile users might be assembled and maintained. Furthermore, as time passes and more data are collected, the opportunity to quantify results with greater confidence will help to solidify best practices for optimizing mobile survey operations, with the caveat that this work is focused on a fast-moving technology where change and evolution is the norm.

Finally, whether we partner with professional developers from the world outside of survey research or train our staff to understand the perspective of other industries, flexibility and agile design that keeps up with growing mobile trends will be critical. Because this medium is growing so rapidly, we should not wait for the perfect study to come along. We must think creatively, embrace experimentation, and be prepared to try and fail. Only through such efforts can we gain a better understanding of this revolutionary tool for survey research.

REFERENCES

American Association for Public Opinion Research Cell Phone Task Force (2010). *New considerations for survey researchers when planning and conducting RDD telephone surveys in the U.S. with respondents reached via cell phone numbers.* Retrieved from http://www.aapor.org/AM/Template.cfm?Section=Cell_Phone_Task_Force_Report&Template=/CM/ContentDisplay.cfm&ContentID=3189.

Apple (2012). *Apple's app store downloads top 25 billion.* Retrieved from http://www.apple.com/pr/library/2012/03/05Apples-App-Store-Downloads-Top-25-Billion.html.

Bardes, B. A., & Oldendick, R. W. (2000). *Public opinion: Measuring the American mind.* Belmont, CA: Wadsworth/Thompson Learning.

Blodget, H. (2012). *Actually, the US smartphone revolution has entered the late innings.* Retrieved from www.businessinsider.com/us-smartphone-market-2012-9.

Blumberg, S., & Luke, J. (2012). *Wireless substitution: Early release estimates from the National Health Interview Survey, January–June 2012.* National Center for

Health Statistics. Retrieved from http://www.cdc.gov/nchs/data/nhis/earlyrelease/wireless201112.htm.

BUSKIRK, T. D., & ANDRUS, C. (2012). Smart surveys for smartphone: Exploring various approaches for conducting online mobile surveys via smartphones. *Survey Practice.* Retrieved from http://surveypractice.wordpress.com/2012/02/21/smart-surveys-for-smart-phones/.

CALLEGARO, M. (2010). Do you know which device your respondent has used to take your online survey? *Survey Practice.* Retrieved from http://surveypractice.wordpress.com/2010/12/08/device-respondent-has-used/.

CALLOWAY, M. (2012). *Cutting through app overload. We media.* Retrieved from http://wemedia.com/2012/02/28/cutting-through-app-overload/.

CHRISTIAN, L., KEETER, S., PURCELL, K., & SMITH, A. (2010). *Assessing the cell phone challenge. Pew Internet & American Life Project, 5/20/2010.* Retrieved from http://www.pewresearch.org/2010/05/20/assessing-the-cell-phone-challenge/.

COUPER, M. P. (2010). *Visual design in online surveys: Learnings for the mobile world.* Presented at the Mobile Research Conference 2010, London. Retrieved from http://www.mobileresearchconference.com/uploads/files/MRC2010_Couper_Keynote.pdf.

DAWSON, A. (2010). *Mobile web design: Best practices.* Retrieved from http://sixrevisions.com/web-development/mobile-web-design-best-practices/.

FREY, J. H. (1983). *Survey research by telephone.* Beverly Hills, CA: Sage Publications.

GIBBS, N. (2012). Your life is fully mobile. *Time Magazine.* Retrieved from http://techland.time.com/2012/08/16/your-life-is-fully-mobile.

GROVES, R. M., FOWLER, F. J., COUPER, M. P., LEPKOWSKI, J. M., SINGER, E., & TOURANGEAU, R. (2004). *Survey methodology.* New York: Wiley.

Internet World Stats (2012). World Internet usage and population statistics: June 30, 2012. Retrieved from http://www.internetworldstats.com/stats.htm.

KAKU, M. (2011). *Physics of the future: How science will shape human destiny and our daily lives by the year 2100.* New York: Knopf Doubleday.

KOCH, P. -P. (2010). Smartphone browser landscape, *A List Apart,* 320. Retrieved from http://www.alistapart.com/articles/smartphone-browser-landscape/.

MCHUGH, M. (2012). *Fighting app fatigue? Crosswa.LK is coming to iPhone. Digital trends.* Retrieved from http://www.digitaltrends.com/mobile/fighting-app-overload-crosswa-lk-comes-to-iphone/.

Nielsen Company, The. (2012). *Nielsen Wire: Nielsen tops of 2012: Digital.* Available at http://blog.nielsen.com/nielsenwire/online_mobile/nielsen-tops-of-2012-digital/.

Online Publisher's Association (2012). *OPA study defines today's smartphone user.* Retrieved from http://www.online-publishers.org/index.php/opa_news/press_release/opa_study_defines_todays_smartphone_user.

PEREZ, S. (2011). *App-ocalypse. Tech crunch.* Retrieved from http://techcrunch.com/2011/12/18/app-ocalypse/.

Pew Research Center (2012). *Smartphone ownership update: September, 2012. Pew Internet & American Life Project, 9/7/2012.* Retrieved from http://pewinternet.org/Reports/2012/Smartphone-Update-Sept-2012.aspx.

PEYTCHEV, A., & HILL, C. A. (2010). Experiments in mobile web survey design: Similarities to other modes and unique considerations. *Social Science Computer Review, 28*(3), 319–335.

Phonescoop. (2012). *Definition of a smartphone.* Retrieved from http://www.phonescoop.com/glossary/term.php?gid=131.

SMITH, A. (2012). *Cell Internet use 2012. Pew Internet & American Life Project, 6/26/2012.* Retrieved from http://pewinternet.org/Reports/2012/cell-internet-use-2012.aspx.

VAUGHAN, P. (2012). *25 Eye-popping Internet marketing statistics for 2012. Hub spot.* Retrieved from http://blog.hubspot.com/blog/tabid/6307/bid/30495/25-Eye-Popping-Internet-Marketing-Statistics-for-2012.aspx.

WELLS, T., BAILY, J., & LINK, M. (2012). *A direct comparison of mobile vs. online survey modes.* Presented at the annual conference of the American Association for Public Opinion Research, Orlando, FL.

CHAPTER 8

Crowdsourcing: A Flexible Method for Innovation, Data Collection, and Analysis in Social Science Research

Michael Keating, Bryan Rhodes, and Ashley Richards,
RTI International

Crowdsourcing is a relatively recent phenomenon that arose in the early part of the 21st century. Jeff Howe first introduced crowdsourcing in a 2006 article in *Wired* magazine, defining it as the act of outsourcing tasks to a large group of people rather than having employees or contractors perform the tasks. The crowd finds out about the task through an open call asking for contributions, and those who respond are typically paid for their time and effort (Govindaraj et al., 2011). Crowdsourcing has also been described as "tapping into the collective intelligence of the public to complete a task" (King, 2009). As Howe originally explained, "it's not outsourcing, it's crowdsourcing" (Howe, 2006).

Social Media, Sociality, and Survey Research, First Edition.
Edited by Craig A. Hill, Elizabeth Dean, and Joe Murphy.
© 2014 John Wiley & Sons, Inc. Published 2014 by John Wiley & Sons, Inc.

WHAT IS CROWDSOURCING?

But what is crowdsourcing exactly? Often the term is thrown around without much context. Social science researchers' experience with crowdsourcing comes in three high-level forms.

- **Open Innovation:** Challenge-based crowdsourcing is used to generate research ideas from a large base of people.
- **Data Collection:** Crowds can be used to conduct targeted data collection to supplement and add depth to social science research. Some researchers are using platforms like Mechanical Turk (discussed below) to gather survey response.
- **Analysis:** Sentiment analysis coding can be conducted on a large scale at relatively low cost using crowdsourcing platforms, and data analysis competitions shed new light on research questions.

The following sections discuss each form of crowdsourcing and give real-life examples of applied crowdsourcing in research.

Crowdsourcing hinges on the idea that individuals in online communities can collectively solve problems and improve research. Perhaps the best-known crowdsourcing network is Amazon Mechanical Turk, an online community that matches temporary workers with paid tasks to perform. Mechanical Turk is based on the idea that humans still perform many tasks better than computers. Each job posted on Mechanical Turk is called a human intelligence task (HIT), and the tasks are generally small. Workers are paid per HIT.

Compared with Mechanical Turk, many crowdsourcing platforms present much larger tasks, often as contests. These platforms include TopCoder and InnoCentive. TopCoder is a network of 250,000 mathematicians, engineers, and software developers worldwide who participate in coding competitions (Warwick & Norris, 2010). InnoCentive is a platform with a broader reach, consisting of 200,000 engineers, scientists, inventors, businesspeople, and research organizations. InnoCentive launched its website in 2001 to help a pharmaceutical maker get ideas from outside the company (Howe, 2006). Like TopCoder, InnoCentive also has a global reach—participants are registered from over 200 countries (Zheng et al., 2011). Crowdsourcing platforms are discussed in the following sections.

Researchers are just beginning to understand the power of crowd-sourcing in social science, and they are excited about its prospects. As shown in the following examples, crowdsourcing methods have the potential to impact research significantly, from brainstorming innovative research ideas to data collection to data analysis.

OPEN INNOVATION

Open innovation is one of the most powerful dimensions of crowdsourcing in social science. An open innovation approach lowers the barriers to entry in research, allows a broad and diverse group of people to give researchers their ideas, and moves beyond a traditional expert panel. Our experience is that open innovation is well suited to help researchers identify gaps in their current research and to take idea generation to a new level of effectiveness.

Open innovation typically takes the form of a challenge or contest. Most crowdsourcing contests do not restrict participation (Archak & Sundararajan, 2009). Allowing open participation is beneficial because participants tend to be successful even in fields in which they have no formal expertise (Howe, 2006). Crowdsourcing contests rarely have an entry fee, but the prizes can be substantial, typically ranging from thousands of dollars to a million dollars (Archak & Sundararajan, 2009), with variation in prizes between platforms or networks. TopCoder winners typically receive $500 to $2,000 (Brandel, 2008), while InnoCentive winners receive $10,000 to $100,000 (Howe, 2006). For example, Netflix's contest to improve the company's movie rating system for a $1,000,000 prize (Daks, 2009) certainly drew attention to crowdsourcing.

The contest prize amount can significantly affect the success of the contest. Higher rewards generally result in more contest entries (Zheng et al., 2011). Archak and Sundararajan suggest allocating the entire budget to the top prize when participants are risk-neutral (i.e., contestants place little importance on prizes), and when participants are risk-averse (i.e., contestants place more importance on prizes), they suggest offering more prizes (2009).

People participate in crowdsourcing contests for various reasons; many are intrinsic, such as the opportunity to learn a new skill or to develop creative skills. Even when the reward in a contest is substantial, research has found that intrinsic motivation remained high among contest participants (Zheng et al., 2011). Other factors driving the motivation to participate include the potential to earn money, the opportunity for participants to market themselves, and the possibility of finding work or a job. Additionally, positive attitudes toward the contest's sponsor may drive participation (Zheng et al., 2011).

The next sections describe two recent applications of open innovation in research project development—the 2009 Cisco challenge and the RTI 2012 Research Challenge.

Cisco Systems I-Prize Challenge

The for-profit sector also uses open innovation methods and challenges to stimulate idea generation. One example is a crowdsourcing contest Cisco sponsored to generate innovations to expand its business. They used a platform from the company, Brightidea, that enabled people to contribute their own ideas and comment and vote on other ideas. The community aspect was a key part of the contest: No anonymous submissions were accepted, and participants created profiles so Cisco and other participants knew who was participating. Many people realized through one another's comments that they had similar ideas and worked together in small teams to develop stronger ideas. Among the ideas that were judged to be finalists, 70% were submitted by teams that had collaborated with others in the online community. One team even consisted of members in California, Singapore, and India (Jouret, 2009).

Few of the 1,200 ideas submitted in Cisco's contest were complete, but all were considered during the process of narrowing down the entries to a final field of 40. Narrowing down the ideas was an extensive task that took six full-time Cisco employees 3 months to complete. When evaluating the submissions, judges asked the following questions:

1. Does this idea address a real pain point?
2. Will it appeal to a sufficiently big market?
3. Is the timing right?
4. If we pursue the idea, will we be good at it?
5. Can we exploit the opportunity for the long term, or would this market commoditize so quickly that we would not be able to stay profitable?

The 40 finalist teams that were selected were then assigned a mentor to help them improve their ideas over the next 6 weeks. After the improvement period, the judges narrowed the ideas to a final field of 10; respective team members were then interviewed via videoconference. The judges were Cisco executives, the contest evaluation team (also Cisco employees), and a Silicon Valley entrepreneur who was invited to give feedback to the teams. These interviews were the last step before the judges selected a winner (Jouret, 2009).

RTI International's 2012 Research Challenge

In 2012, RTI International developed plans to conduct an omnibus survey of people in Chicago, IL, to test a variety of new survey research methods. As our team began questionnaire development, we decided to open up the activity and crowdsource the survey content to offer it as a resource to address a wide range of research questions. So, we asked social science researchers to nominate topics and questions on social, economic, and health issues.

Any researcher was invited to submit an entry to the challenge. The team encouraged people to submit entries that addressed creative, cutting-edge, and pressing social, economic, or policy topics. Submissions were welcome on topics from education, health, and economics to sociology, political science, energy, and the environment. The team asked researchers to prepare and submit a set of 10 or fewer survey questions. The questions could be open- or close-ended. The team also asked those submitting questions to include a two-page synopsis that outlined and explained the dimensions of their research project (see Figures 8.1 and 8.2). These dimensions included:

- **Introduction/Nature of the Problem:** We asked researchers to describe briefly the issue or problem on which they sought information. This issue or problem was based on previous research or literature, but a full literature review was not required.
- **Research Question:** We asked researchers to specify the research question that their survey questions sought to answer.
- **Justification:** We asked researchers to explain why and how their research would advance the state of knowledge about the problem.
- **Dissemination Plan:** We asked researchers to describe plans for preparing manuscripts, publishing their study findings, and/or giving conference presentations.

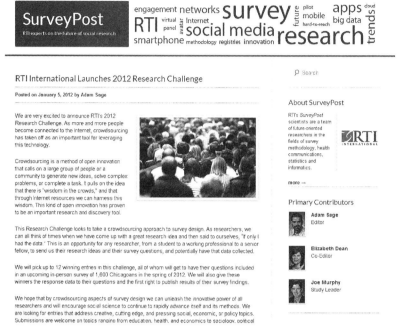

FIGURE 8.1 Screenshot of RTI Research Challenge rules page.

As mentioned earlier in this chapter, money is often used to motivate challenge participants; however, the team decided to use nonmonetary incentives that were appropriate for the target contest entrants. Participants were told that winners would have their set of questions included

FIGURE 8.2 Word cloud of 2012 Research Challenge entries.

TABLE 8.1
Employment Background of Participants

Area of Employment	Percentage of Participants	Percentage of Award Winners
Academia	66.1%	22.2%
Private Sector	13.6%	—
Nonprofit	1.7%	—
Government	5.1%	11.1%
Student	25.4%	66.7%
Other	5.1%	—

in the survey, and the winning researchers would receive the response data to their questions along with demographic data and the first rights to publish results of their survey findings.

The challenge lasted for 25 days, and the team received 74 entries that covered many topics from use of emerging tobacco products to local perceptions of law enforcement. As shown in Table 8.1, participants in the contest came from a diverse set of backgrounds in academia, government, and the private sector.

Judging was blind, and we selected nine winners. Of these nine, six were graduate students who almost certainly would have been excluded from a traditional expert panel. This result is particularly stunning, considering that students were only 25.4% of the pool of entrants. One hundred percent of participants surveyed said they would consider entering a similar challenge in the future.

Options for Hosting Your Own Challenges

With the rise of crowdsourcing challenges and contests, online platforms have emerged to host and publicize the events. ChallengePost (www.challengepost.com) is a company that allows users to host their challenges on the website on a broad range of topics including education, business, environment, and media. Their community includes over 300,000 users. ChallengePost will actively promote and manage challenges for businesses and organizations. They also allow potential participants to create discussions and brainstorm ideas to enhance collaboration.

Challenge.gov (www.challenge.gov) is the U.S. government's challenge platform that was created in reaction to the 2009 Strategy for

American Innovation, which encouraged agencies to use crowdsourcing methodologies to promote innovation. Diverse government agencies use this platform to post challenges, including the U.S. Department of Defense, the National Aeronautics and Space Administration, the U.S. Department of Labor, the Centers for Disease Control and Prevention, and the U.S. Environmental Protection Agency. These challenges range in compensation, and some prizes can be as high as $500,000.

InnoCentive (www.innocentive.com) is a challenge hosting and management company similar to ChallengePost, but its challenges tend to focus much more on the hard sciences like chemistry, engineering, and agriculture. Their pool of "registered solvers" includes more than 270,000 people in countries worldwide. In addition, InnoCentive states that their service can reach more than 12 million people using their trusted partners. Public and private organizations regularly post challenges and competitions on this site.

Although these platforms are useful and convenient, researchers can also host and manage challenges on their own. Depending on the research topic, it is generally easy to target professional e-mail lists and social media groups that allow a large number of relevant people to view your challenge.

Legal Considerations

Protecting participants from intellectual property theft is a major concern in crowdsourcing contests (Brandel, 2008), and the Cisco contest was no different. Developing the legal framework for the Cisco contest was challenging, especially because the company did not want to discourage or intimidate prospective participants with too much legalese. When they submitted an idea, participants agreed to pledge that the idea was theirs, not someone else's, and if they won the contest, they would give Cisco the rights to their idea in exchange for $250,000. The agreement also stipulated that Cisco would not own the nonwinning ideas submitted as part of the contest. As a precaution, only a small team of judges saw the ideas submitted in the contest (Jouret, 2009). InnoCentive takes similar precautions in its contests, allowing only the organization that posted a problem to see the submitted ideas (Brandel, 2008). In addition, Cisco's legal framework also addressed how Cisco would be protected if a participant submitted an idea that Cisco was already working on internally.

In addition to protecting participants, organizations considering launching a challenge should develop a legal framework to protect themselves, as RTI did when developing the 2012 Research Challenge. The terms and conditions in the RTI Challenge agreement clearly stated that if a participant submitted an entry that infringed on the copyright or rights of a third party, the participant was legally responsible for any damages or legal costs.

Challenge-based approaches hold much promise for researchers. Traditionally, researchers and businesses assemble a small panel or group of experts to generate ideas and approaches. Unfortunately, for people with less experience, this approach creates relatively high barriers to entry in idea generation and innovation. Open innovation approaches seek to eliminate those barriers and to engage all researchers at all levels. The results of the RTI 2012 Research Challenge suggest that when these barriers are removed, graduate students can compete with the best minds and experts in a field.

DATA COLLECTION

Approaches to data collection are shifting as researchers are realizing the potential of using crowds to collect data and to respond to surveys. This section examines four types of crowdsourced data collection that are gaining popularity in social science research.

1. Crowdsourcing survey response
2. Targeted data collection efforts
3. Challenge-based data collection
4. Citizen observation networks

Each of these approaches to data collection varies in its methods of motivating individual behavior in crowds and gathering responses. Characteristics that cut across all types of crowdsourced data collection include lowered costs and speed. However, researchers need to consider numerous issues if they plan to use crowdsourcing data collection methods to ensure that high-quality, accurate data are collected.

Crowdsourcing Survey Response on Mechanical Turk

As more researchers turn to Mechanical Turk to collect survey data, others question how workers on Mechanical Turk compare with the greater

TABLE 8.2

Location of Mechanical Turk Workers

Country	Percentage
United States	46.8%
India	34.0%
Other	19.2%

Source: Ipeirotis, 2010.

population. A survey of 1,000 Mechanical Turk workers conducted in February 2010 found the following:

- Of total participants, 47% were from the United States, 34% were from India, and 19% were from other countries (see Table 8.2).
- In the United States, significantly more females than males participated, while the opposite was true in India (see Table 8.3).
- The age of participants skewed toward younger workers, especially in India.
- Participants had a higher education level and a lower income than the general U.S. population.
- Most participants did not have children, and a high number were single.
- Most participants indicated that they earned less than $20 per week through Mechanical Turk, but a small number earned more than $1,000 per month (Ipeirotis, 2010).

Despite these differences, Mechanical Turk survey participants were generally more demographically representative than samples of undergraduates typically used in university studies (Bohannon, 2011). One

TABLE 8.3

Gender Breakdown of U.S. Mechanical Turk Workers

	Mechanical Turk	General Population
Female	65.0%	50.9%
Male	35.0%	49.1%

Sources: Ipeirotis, 2010; U.S. Census, 2010.

study, which compared results of surveys given with a Mechanical Turk sample and a typical university research sample of undergraduates, found that the Mechanical Turk sample was older, more likely to be employed, and more diverse in education, employment status, and profession. Both the university sample and the Mechanical Turk sample were primarily female and Caucasian, but overall, the Mechanical Turk sample more closely resembled the greater population (Behrend et al., 2011).

In the same study, researchers compared the quality of responses given from the Mechanical Turk sample and the university sample. They calculated Cronbach's coefficient alpha and found higher internal consistency for the Mechanical Turk sample; however, using a 33-point scale, they discovered higher social desirability scores too. They found no differences between the groups in completion time or word count. Finally, their results showed that the Mechanical Turk sample had more computer and Internet knowledge (Behrend et al., 2011).

Compared with a university sample, a Mechanical Turk sample seems to be more diverse. Another benefit of conducting a survey with Mechanical Turk is the quick turnaround time. Behrend et al. recruited several hundred Mechanical Turk participants in less than 48 hours. Participants in the Mechanical Turk sample were paid $0.80 for the 30-minute survey, which was slightly above the median payment for a 30-minute task (Behrend et al., 2011). One drawback of conducting a survey this way is that if participants do not read the informed consent carefully, they may feel pressured to complete a Mechanical Turk survey because they do not understand that it is not tied to their rating or compensation (Behrend et al., 2011).

Other researchers have used Mechanical Turk as a malaria surveillance tool (Chunara et al., 2012). Because a large proportion of Mechanical Turk workers live in India, the platform was conducive to studying malaria prevalence in that country. The researchers argue that with careful design, micro-monetary incentive self-reporting studies can be used for public health surveillance.

Using Mechanical Turk as compared with a university sample certainly has advantages, including a more diverse pool of participants, rapid data collection, and low cost. Researchers can also isolate Mechanical Turk responses to specific countries. However, Mechanical Turk does not currently offer the ability to dig down further to smaller geographic areas. Thus, although studies like the malaria surveillance study can request Mechanical Turk participants from specific areas (e.g., Mumbai) in the HIT title, no enforcement mechanism prevents workers in other parts of India from taking the survey. This limitation is a significant weakness in the survey design.

Researchers may consider using Mechanical Turk to crowdsource their survey response for several reasons, including low cost, timeliness, and convenience. However, Mechanical Turk is not a replacement for a traditional random sample. Additionally, it is difficult or impossible to verify that the responses from Mechanical Turk workers are truthful. Although a large-scale representative data collection effort is not yet possible on Mechanical Turk, the possibility of carrying out social science experiments on the platform has many researchers excited (Frick, 2012). It is in this experimentation and pretesting arena that Mechanical Turk holds the most potential for survey response.

Targeted Data Collection

Targeted data collection can be used to create unique and specific sets of data for researchers. Crowdsourcing this type of targeted data collection can be used to supplement traditional social science research data. The targeted data collection examples discussed in this section show the potential of using crowds to add new depth to social science research that was not possible a decade ago.

Emerging Tobacco Product Detection

One of the winners of RTI's 2012 Research Challenge proposed to examine the demand-side dynamics of snus smokeless tobacco products. Snus is part of a new genre of smokeless tobacco products designed to be more appealing compared with traditional smokeless tobacco products because it does not require spitting (Pechacek, 2010).

These dynamics were to be captured in survey questions that gauged participants' knowledge of and potential interest in using these newer tobacco products. However, one dimension not captured in a survey is the supply of these smokeless tobacco products. The team asked a simple question, "Where is snus sold in Chicago?" The rationale was that concentrations of tobacco retailers that sell snus tobacco products could influence the demand-side dynamics of the nearby local population.

To answer this question, the team obtained a publicly available list of registered tobacco retailers in Chicago from the city government.

The team used this list to develop and post HITs on Mechanical Turk that asked workers to call retailers and inquire whether they sold snus tobacco products. Workers would then record the outcome of the call and submit the HIT for review (see Figure 8.3). Within a few hours of launching these HITs on Mechanical Turk, the team had assembled an entirely new micro-level dataset of snus tobacco in an urban area.

The data collected using these methods were compiled into a map for visualization purposes (see Figure 8.4). In addition to supplementing survey data, this retailer data can be combined with other public data to understand potential marketing and supply approaches of tobacco companies.

Findings from the Snus Study

Collecting data using Mechanical Turk is *extremely* fast. For example, one step in the snus data collection process was determining the phone numbers of registered tobacco retailers on the public list. One phone number data collection batches included 2,053 HITs, and data collection was completed within 2 hours. The same work could have taken days (rather than hours) to complete using more traditional methods. The speed of this data collection approach allows researchers to obtain data in close to real time. Findings also showed that Mechanical Turk workers reached a consensus on the correct phone number in 97% of cases. If researchers need to conduct data collection in reaction to a time-sensitive event, Mechanical Turk may be a useful option.

Data quality is a huge concern when using a decentralized microtask workforce like Mechanical Turk. The quality assurance team should be able to verify data collection tasks easily. Data that are not verifiable cannot be assumed to be reliable and valid, regardless of the method of collection.

Although Mechanical Turk users can prescreen workers for a number of completed HITs and percentage approved (all of which are considered metrics of quality by Amazon), falsification does happen. RTI used several methods to ensure that data collected were valid, which can be applied to other research studies that use Mechanical Turk.

First, multiple workers should always complete the same HIT. By having multiple workers complete the same task, researchers can analyze the reliability of the data collection results to check for falsification. For example, one worker has a 50% chance of randomly selecting the correct

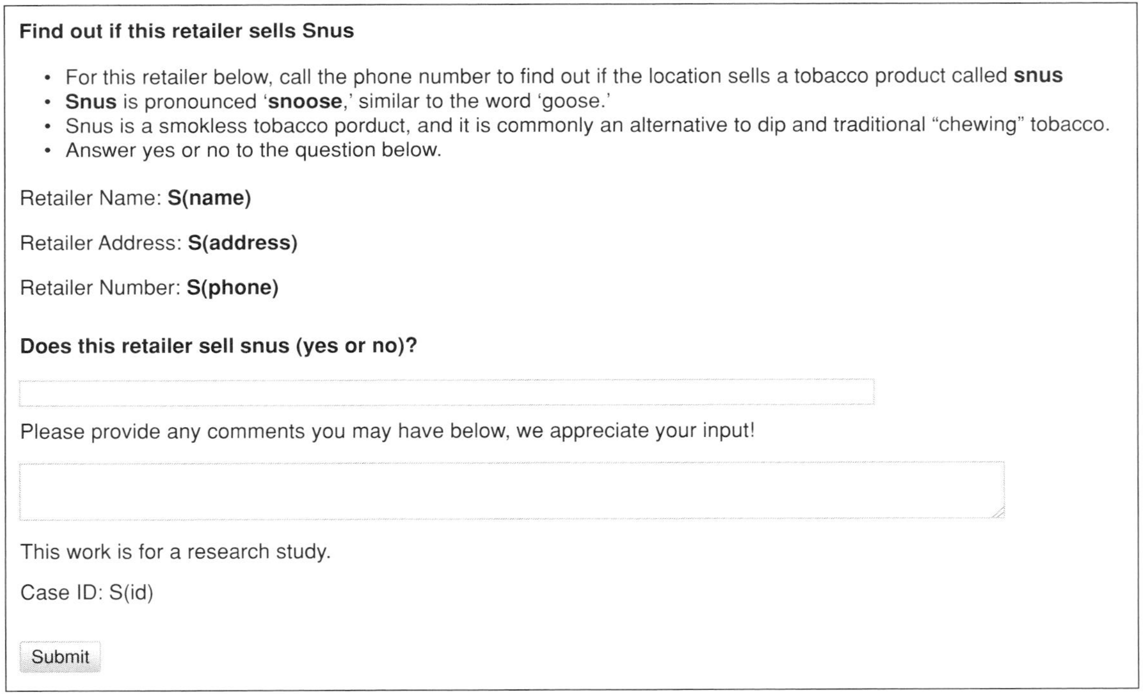

Find out if this retailer sells Snus

- For this retailer below, call the phone number to find out if the location sells a tobacco product called **snus**
- **Snus** is pronounced '**snoose**,' similar to the word 'goose.'
- Snus is a smokeless tobacco porduct, and it is commonly an alternative to dip and traditional "chewing" tobacco.
- Answer yes or no to the question below.

Retailer Name: **S(name)**

Retailer Address: **S(address)**

Retailer Number: **S(phone)**

Does this retailer sell snus (yes or no)?

Please provide any comments you may have below, we appreciate your input!

This work is for a research study.

Case ID: S(id)

Submit

FIGURE 8.3 Screenshot of Amazon Mechanical Turk hit template for snus data collection. © 2005–2013 Amazon.com, Inc. or its Affiliates.

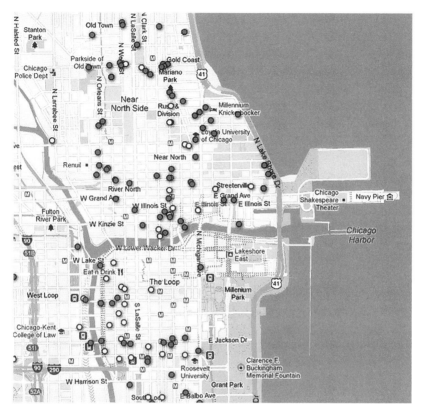

FIGURE 8.4 Map of tobacco retailers that sell snus in various Chicago zip codes. *Note:* Gray circles indicate that snus is not sold at that location, while white circles indicate that snus is sold at that location. © 2013 Google – Map data © 2013 Google, Sanborn.

answer to a simple Yes or No question through falsification. If five independent workers perform the same task, the probability of all five workers randomly selecting the same answer is 3.1% (3.1% = .5 × .5 × .5 × .5 × .5). Before beginning a data collection effort on Mechanical Turk, always consider the acceptable probability of falsification.

Second, researchers should conduct analysis on all of their workers to determine how often each worker agrees with a majority of workers doing the same tasks. By doing this quick analysis, research teams can quickly filter out workers who are falsifying. If a worker agrees with others only 10% of the time, then there is a high probability that falsifying or another problem is occurring.

Third, researchers should embed dummy cases to conduct quality control of specific workers. This step allows research teams to pinpoint specific workers of concern that other analyses may miss. For example, suppose a research team is undertaking a rapid telephone data collection effort involving retail stores. Based on experience, RTI recommends that this team include phone numbers that look like real cases but actually connect to people on the research team. If the research team never receives a call from a Mechanical Turk worker, but the worker somehow completes an HIT, then research can identify falsification. This research team could then reject those workers' HITs, block the workers, and repost the HITs for someone else.

Finally, researchers should consider having trusted team members complete a certain percentage of HITs to determine outcomes independent of the Mechanical Turk workers. This step can also help determine whether some workers are providing bad data. Most Mechanical Turk workers are honest; however, anyone using the service should closely monitor the quality of their data. Quality control is a best practice; no one wants results to be compromised by a few errant or dishonest workers.

Cost Considerations

Low cost is one commonly cited benefit of using platforms such as Mechanical Turk. For example, rather than paying participants US$8 for a 15–20-minute experiment in a lab, a political scientist at the Massachusetts Institute of Technology in Cambridge pays a Mechanical Turk participant US$0.75 to US$1.00. That rate is roughly comparable with the compensation in a political science study conducted by researchers at the University of California at Berkeley, which paid participants in the United States US$0.20 to US$0.25 and participants in India US$0.15 for completing a 4-minute task (Bohannon, 2011). Another example is a survey that took about 3 minutes to complete and paid participants US$0.10 (Ipeirotis, 2010). Payment for tasks is especially important: If it is too low, workers may not be interested, but if it is too high, it may attract spammers who will do low-quality work. One suggestion is to pay US$0.15 to US$0.50 for a 10-minute job on Mechanical Turk (Bohannon, 2011), although some tasks pay significantly less (e.g., US$0.11 for taking a 30-minute survey) (Kapelner & Chandler, 2010).

Researchers should be strongly encouraged to pay workers an adequate wage when they use platforms like Mechanical Turk. If one assumes that participants in Bohannon's study could complete 15 tasks taking 4 minutes each in an hour for US$0.25 per task, then workers' hourly wages would only be US$3.75. This wage is significantly below the minimum wage in the United States, and it raises serious ethical concerns. The claim that higher wages may attract spammers may have merit; however, as discussed in the data quality section, researchers can and should take steps to filter out spammers. Amazon does not charge users for rejected work, so spammers are detected and the customer/researcher does not lose money. During data collection from snus tobacco retailers, workers were paid US$0.75 per HIT and the team assumed that on average the task would take about 4 minutes to complete. If a worker completed 15 HITs in 1 hour, his hourly wage would be US$11.25. RTI's team encountered some spammers, but their work was rejected and they were not compensated.

One of Amazon Mechanical Turk's efficiencies is that researchers can rapidly scale up and scale down their workforce to meet their data collection needs. Hundreds of data collectors can begin working on HITs within minutes of launch, and when the HITS are complete, workers are owed only financial compensation for their work. The speed of this workforce scaling process is very appealing.

MyHeartMap Challenge

Other researchers have undertaken a challenge-based targeted data collection approach. In 2012, researchers at the University of Pennsylvania launched the MyHeartMap Challenge (http://www.med .upenn.edu/myheartmap). This challenge sought to find and photograph as many automated external defibrillators (AEDs) in Philadelphia, PA, as possible. Researchers conducted the challenge for 2 months, and more than 300 individuals and teams participated, finding more than 1,500 AEDs in 800 unique buildings around the city (http:// www.uphs.upenn.edu/news/News_Releases/2012/05/myheartmap/).

This challenge used a variety of monetary incentives to motivate people to participate. In the end, two individuals each received a US$9,000 grand prize for finding over 400 AEDs each. The research team deemed a specific number of AEDs in Philadelphia as "Golden" AEDs, and if participants found the AED, they were awarded a US$50 prize. This varied incentive approach was intended to keep participants engaged throughout the challenge, and this aspect of gamification cannot be underestimated (see Chapter 11 in this volume for more information on gamification).

Crowdsourced Citizen Observation Networks: eBird and Waze

Technological advances have led to the rise of an entirely new, dispersed kind of data collection. Entire citizen science networks have sprung up that organically create a crowd of individuals with a common interest. Researchers can use technology networks to gather crowdsourced data: Two examples include eBird and Waze.

eBird (www.ebird.org) allows users to record birds, explore dynamic data on bird sightings and migrations, and share sightings with a community of bird watchers. This type of data collection is changing the way researchers study biology and conservation (Sullivan et al., 2009). Collectively, participants in these networks amass a large quantity of invaluable data for researchers. For example, as of March 2012, eBird claims to have more than 3.1 million bird observations on record (http://ebird.org/content/ebird/about). At any moment scientists can see real-time data from 180,000 locations around the globe (Sullivan et al., 2009). These data allow scientists to observe seasonal distribution changes in bird locations, variations in bird migration range, and differential timing in migration patterns.

Other data collection platforms like Waze allow users to post observations of traffic incidents, for example, road construction, hazards, or accidents. Taken collectively, these observations give users an idea of the local traffic situations surrounding them. Companies like Google are also using anonymized global positioning system (GPS) data to determine trip durations based on location (Google, 2009). Collecting GPS data from individuals periodically allows Google to determine traffic conditions on roads around the United States and to provide a traffic layer in Google Maps for all consumers (see Figure 8.5).

The citizen observation approach to crowdsourced data collection is unique in that participants are incentivized by self-motivation alone. Many participants in these networks are concerned about the topic or the issue is important to them. Collections of individuals gathering data now allow researchers to see near real-time trends. Compared with traditional data collection, the costs of these methods tend to be much lower.

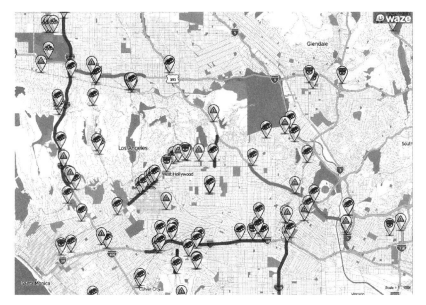

FIGURE 8.5   Screenshot of Waze traffic updates. © 2013 Waze Mobile Ltd.

ANALYSIS BY CROWDSOURCING

After leveraging crowds to design a research study or using novel data collection methods, analysis is the logical next step. Sentiment analysis and data analysis are two rapidly growing areas of crowdsourcing analysis.

Sentiment Analysis

As the collection of social media data continues to gain popularity, researchers increasingly have large amounts of unstructured data that need to be analyzed. One method is sentiment analysis. Currently, humans can code social media data like Tweets much more accurately than computers (Richards, 2012). And crowdsourcing platforms give researchers the ability to code social media quickly on a large scale.

One such platform is the Mechanical Turk Sentiment Analysis application, which allows researchers to post large batches of social media data for humans to read and code. The application allows requestors to define specifically each sentiment response category to reduce coder error and improve the accuracy of responses.

FIGURE 8.6   Screenshot of U.S. Census return rate challenge on Kaggle. *Source:* Census Bureau.

Using humans to conduct sentiment analysis has several advantages. First, researchers get multiple perspectives on the same item. By having multiple people code the same data, researchers can see a consensus regarding sentiment. Second, humans are more adept than computers at understanding sarcasm and nuance in language (see Chapter 2 in this volume for more detailed information on sentiment analysis in survey research).

Potential Mechanical Turk users should read the Findings from the Snus Study section of this chapter before using Mechanical Turk for analysis. Users must take precautions to ensure data quality. Mechanical Turk also recommends that users of the Sentiment Analysis app have a minimum of five people code the same item.

Challenge-Based Data Analysis

In addition to sentiment analysis, researchers are beginning to use crowds to analyze their data. Platforms like Kaggle (http://www .kaggle.com) allow users to host data analysis competitions. Figure 8.6 is a screenshot of a 2012 U.S. Census competition that asked the crowd to predict and visualize census mail return rates. By the end of the visualization competition, a total of 244 teams were competing for a US$14,000 prize. The strength of platforms like Kaggle is that they bring together diverse people with common interests, in this case quantitative science, to solve problems collectively.

Researchers have also used Mechanical Turk to add to their data analysis through crowdsourced brainstorming approaches (Willett et al. 2012). Researchers posted pictures of data visualizations on Mechanical Turk and asked workers to analyze and annotate or explain different dimensions of the visualizations. This brainstorming approach yielded diverse high-quality results and novel insights (Willett et al., 2012). Using Mechanical Turk workers, researchers can gain a vast amount of information in a short period of time.

CONCLUSION

Crowdsourcing is gaining significant traction in social science circles, and it is already yielding results that should help to advance research at a pace unheard of in the past. Dimensions of crowdsourcing that are particularly appealing are:

- **Diversity in points of view:** Instead of a small team or a panel of experts completing a task, crowdsourcing allows many people to contribute. When generating research ideas, diversity can yield innovative, unexpected results, and when conducting analysis, people from different backgrounds can offer new perspectives on the data.

- **Lowers cost:** Typically crowdsourcing lowers costs by streamlining tasks and removing red tape. Such reduction is particularly evident in the use of Mechanical Turk, which allows researchers to scale their workforce rapidly depending on their immediate needs.

- **Fast results:** Data collection using platforms like Mechanical Turk is extremely fast. During the collection of snus tobacco information, 1,500 cases were completed within 8 hours of posting the HITs online. This rapid turnaround allows researchers to respond quickly to potentially time-sensitive events in their data collection approach.

- **High quality:** If conducted properly, crowdsourcing can yield high-quality results; however, researchers should be encouraged to develop data quality protocols as they design their projects. On Mechanical Turk, many workers are honest, but those who are dishonest may compromise the data (or present a risk of compromise).

Although it is an exciting technology, crowdsourcing has limits. At this point, it is not a suitable replacement for a traditional random sample. Dreams of low-cost representative surveys with high response rates are far from being realized. Crowdsourcing is also an inappropriate choice for some projects. Before researchers decide to pursue a crowdsourcing method, they must consider both the project's goals and from which crowd they wish to solicit input. Nonetheless, crowdsourcing is a useful and potentially revolutionary way to complement survey datasets. The prospect of crowdsourcing becoming a standard social science approach in research is intriguing. Crowdsourcing has the potential to benefit researchers on many different dimensions, including diversification of points of view, cost savings, speed, and quality.

REFERENCES

ARCHAK, N., & SUNDARARAJAN, A. (2009). Optimal design of crowdsourcing contests. *International Conference on Information Systems (ICIS) 2009 Proceedings.*

BEHREND, T. S., SHAREK, D. J., MEADE, A. W., & WIEBE, E. N. (2011). The viability of crowdsourcing for survey research. *Behavior Research Methods, 43*(3), 800–813.

BOHANNON, J. (2011). Social science for pennies. *Science, 334*(6054), 307–307.

BRANDEL, M. (2008). Crowdsourcing: Are you ready to ask the world for answers? *Computerworld, 42*(10), 24–26.

CHUNARA, R., CHHAYA, V., BANE, S., MEKARU, S., CHAN, E., FREIFELD, C., & BROWNSTEIN, J. (2012). Online reporting for malaria surveillance using micro-monetary incentives, in urban India 2010-2011. *Malaria Journal, 11*(43), 1475–2875. Retrieved from http://www.malariajournal.com/content/11/1/43.

DAKS, M. C. (2009). Business solution that won't get lost in the crowd. *Njbiz, 22*(42), 1–12.

FRICK, W. (2012). Mechanical Turk and the limits of big data. *MIT Technology Review.* Retrieved from http://www.technologyreview.com/view/506731/mechanical-turk-and-the-limits-of-big-data/.

Google (2009). *The bright side of sitting in traffic.* Retrieved from http://googleblog.blogspot.com/2009/08/bright-side-of-sitting-in-traffic.html.

GOVINDARAJ, D., NAIDU, K. V. M., NANDI, A., NARLIKAR, G., & POOSALA, V. (2011). MoneyBee: Towards enabling a ubiquitous, efficient, and easy-to-use mobile crowdsourcing service in the emerging market. *Bell Labs Technical Journal*, *15*(4), 79–92.

HOWE, J. (2006, June). The rise of crowdsourcing. *Wired*, (14.06). Retrieved from http://www.wired.com/wired/archive/14.06/crowds.html.

IPEIROTIS, P. (2010). Demographics of Mechanical Turk. *Center for Digital Economy,* Research Working Papers, 10.

JOURET, G. (2009). Inside Cisco's search for the next big idea. *Harvard Business Review*, *87*(9), 43–45.

KAPELNER, A., & CHANDLER, D. (2010, October). *Preventing satisficing in online surveys: A "kapcha" to ensure higher data quality*. Presented at CrowdConf 2010, San Francisco, CA. Retrieved from http://www.crowdsourcing.org/document/preventing-satisficing-in-online-surveys-a-kapcha-to-ensure-higher-quality-data/2538.

KING, S. (2009, January 27). Using social media and crowd-sourcing for quick and simple market research. *U.S. News and World Report*. Retrieved from http://money.usnews.com/money/blogs/outside-voices-small-business/2009/01/27/using-social-media-and-crowd-sourcing-for-quick-and-simple-market-research.

PECHACEK, T. F. (2010). *Smokeless tobacco: Impact on the health of our Nation's youth and use in major league baseball*. Centers for Disease Control and Prevention Congressional Testimony on Energy and Commerce, United States House of Representatives. Retrieved from http://www.cdc.gov/washington/testimony/2010/t20100414.htm.

RICHARDS, A. (2012). *Review of IBM SPSS text analytics for surveys*. Retrieved from https://blogs.rti.org/surveypost/2012/02/17/review-of-ibm-spss-text-analytics-for-surveys/

SULLIVAN, B. L., WOOD, C. L., ILIFF, M. J., BONNEY, R. E., FINK, D., & KELLING, S. (2009). eBird: a citizen-based bird observation network in the biological sciences. *Biological Conservation*, *142*, 2282–2292.

WARWICK, G., & NORRIS, G. (2010). Crowd control. *Aviation Week & Space Technology*, *172*(40), 75–75.

WILLETT, W., HEER, J., & AGRAWALA, M. (2012). *Strategies for crowdsourcing social data* analysis. Retrieved from http://vis.berkeley.edu/ papers/CrowdAnalytics-CHI2012(Preprint).pdf.

ZHENG, H., LI, D., & HOU, W. (2011). Task design, motivation, and participation in crowdsourcing contests. *International Journal of Electronic Commerce*, *15*(4), 57–88.

CHAPTER 9

Collecting Diary Data on Twitter

Ashley Richards, Elizabeth Dean, and Sarah Cook,
RTI International

Twitter is a social media tool that allows users to send and receive quick, frequent messages—called Tweets—of 140 characters or less. Tweets are posted to a user's Twitter profile page, can be linked to blogs and other social media profiles (such as Facebook and LinkedIn pages), and are searchable within Twitter's search engine. Twitter describes itself as an information network rather than as a social media platform because the critical element of the Twitter experience is not the user's profile or connections, but the information contained in each Tweet (Twitter, 2012a). Although many Twitter users are individuals, businesses, organizations, and media increasingly use Twitter accounts for information dissemination and product sharing.

Active Twitter users submit regular updates on Twitter about what is happening in their lives. Tweets often describe ordinary occurrences such as what people ate for breakfast or what movie they saw last night. This kind of daily, habitual reporting is similar to the type of information that researchers aim to collect in diary studies—surveys characterized by frequent self-report of the events and experiences in people's lives. In this chapter, we (1) describe the potential, benefits, and drawbacks

Social Media, Sociality, and Survey Research, First Edition.
Edited by Craig A. Hill, Elizabeth Dean, and Joe Murphy.
© 2014 John Wiley & Sons, Inc. Published 2014 by John Wiley & Sons, Inc.

of collecting diary data on Twitter; (2) describe the process we used to collect data on Twitter in a pilot diary study; and (3) report the findings from that study.

BACKGROUND

Twitter

Twitter has over 140 million active users who send an average of 340 million Tweets per day (Twitter, 2012b). In addition to these active users, a recent study suggested that over 500 million total Twitter accounts now exist (Semiocast, 2012), including infrequent users and fake accounts (Taylor, 2012). In the United States alone, as of February 2012, an estimated 15% of online adults use Twitter, and 8% of online adults use Twitter every day. African American (28%), young (18- to 29-year-olds, 26%), urban (10%), and suburban (14%) online adults are more likely to use Twitter than other demographic groups. One of the best correlates of Twitter use is smartphone ownership. Compared with 9% (ever) and 3% (typical day) for basic mobile phone users, 20% of smartphone users have used Twitter and 13% use it on a typical day (Smith & Brenner, 2012). Overall, 9% of all mobile phone users use Twitter on their phones, with 16% of smartphone users using Twitter on their phones. Among cell owners, rates of Twitter use on cell phones is especially high among 18- to 24-year-olds (22%), non-Hispanic blacks (17%), and Hispanics (12%).

Diaries

A diary is a self-report instrument used to collect frequent reports on the events and experiences in people's lives (Bolger et al., 2003). Diaries can be used for a range of topics such as food or alcohol consumption, tobacco use, transportation, physical activity, time use, and expenditures. Over decades of use for data collection, diaries have taken on many different forms to better exploit available technologies including pagers, handheld computers, and smartphones.

For over a century, paper-and-pencil was the predominant mode for diaries. Topics included how schoolchildren spent their free time (Fox, 1934), to which radio stations people listened (Beville, 1949), drinking habits (Williams & Strauss, 1950), and consumer behaviors (Sudman,

1964). In 1972, the National Consumer Expenditure Survey (now the Consumer Expenditure Survey) began using a diary format to reduce measurement error in reporting expenditures.

Efforts to improve the regularity with which participants complete diary surveys have resulted in increasing technological sophistication of diary data collection tools. Previous diary methods have evolved to incorporate daily telephone calls, pagers for signaling response times and for recording responses (Shrier et al., 2005), interactive voice response (IVR) for receiving inbound diary submissions via dialing a toll-free number (Mundt et al., 1995), two-way text messaging for asthma management (Anhoj & Moldrup, 2004) and measuring pain (Alfven, 2010), and personal digital assistants (PDAs) and smartphone applications (apps) for monitoring dietary intake (Sevick et al., 2010) and pain (Jamison et al., 2001) among other topics.

Smartphones are capable of running apps—programs that have a purpose unrelated to a phone's standard features (e.g., making and receiving calls). Apps and other features of cell phones have provided new options for diary data collection. One example is a photo-ethnographic project that used the built-in cameras on phones (Haworth, 2010). Participants' cell phones were preprogrammed to ring at eight random times per day, at which point participants were asked to take a photo of their surroundings using the phone.

A GPS (global positioning system) is another cell phone feature that can benefit diary studies. One study that took advantage of this feature used cell phones with GPS to track the travel behavior of adolescents (Wiehe et al., 2008). A GPS has been used in other ways as well. For instance, in a study of environmental exposures, participants wore a GPS unit to supplement a paper-and-pencil diary (Phillips et al., 2001).

Diaries on Twitter

Technological advances—particularly cell phones—have facilitated the collection of richer diary data. Twitter is a natural platform for this type of data collection because many users already view Twitter as a diary or record of their lives (Marwick & Boyd, 2010), and Twitter's communication norms support its use for frequently broadcasting announcements about what users are doing and how they are feeling. Furthermore, Twitter's popularity with smartphone users and its accessibility make Twitter a tool worth considering for diary survey data collection.

Reporting diary data via Twitter on a personal mobile device may reduce the burden for participants and increase compliance because many people Tweet as part of their usual routine. They already know how to use Twitter and would not add a new component to their daily life when they participate in a study that uses it. Forgetting to check Twitter is a problem that can also be remedied as Twitter.com and other user agents will send notifications, either on a mobile device or through e-mail, when users have a new message.

Data quality may benefit from the use of Twitter via mobile devices because entries can be inserted closer to the time of origin and can be supplemented with additional sources of rich data, such as photographs. Furthermore, data quality may improve for certain study topics because Twitter encourages a fast-paced, extemporaneous style of communication and participants may be more inclined to respond without over-thinking what they report.

Twitter can also benefit diary studies because, as a social network, Twitter encourages users to share information. Participants in a study of nonsensitive behaviors may be more likely to enter data by sharing their information on Twitter. Furthermore, seeing that other participants have responded could serve as a reminder and motivation to respond.

METHODS

As part of our exploratory study, we tested a number of different methods by conducting six diary studies with small, nonprobability samples. Recruiting, screening, and coordinating with participants involved multiple steps and interfaces.

First, we set up seven Twitter accounts. Twitter allows only one account per e-mail address, so we also set up seven e-mail addresses to create the accounts. We created one account for each of the six diary designs we planned to test and one extra account for general study recruitment purposes. Three diary accounts used topic-specific names (RTIFoodSurvey, RTIMoodSurvey, and RTITimeSurvey), and three used general names (RTISurvey2, RTISurvey5, and RTISurvey6). The general names are advantageous because they are concise and use up less of the character limit for a Tweet; however, the names are less related to their respective studies and can easily be mistyped.

The diaries varied on the following factors: target population (young adults ages 18 to 24, Hispanics, people with diabetes), diary duration

TABLE 9.1
Study Design

	RTI FoodSurvey ($n = 9$)	RTI Survey2 ($n = 13$)	RTI MoodSurvey ($n = 9$)	RTI TimeSurvey ($n = 12$)	RTI Survey5 ($n = 11$)	RTI Survey6 ($n = 15$)
Topic	Diet	Activity	Mood	Activity	Diet	Mood
Sample	Diabetes	Diabetes	Hispanic	Hispanic	Young Adult	Young Adult
Duration	1 week	4 weeks	3 days	1 week	3 days	4 weeks
Incentive	$30 (Guaranteed)	Up to $30 ($2/Tweet)	Up to $20 ($1/Tweet)	Up to $10 ($1/Tweet)	$10 (Guaranteed)	$20 (Guaranteed)

(3 days, 1 week, 4 weeks), diary topic (mood, diet, activity), incentive amount ($10, $20, $30), incentive form (Amazon.com gift card, choice of Amazon.com or iTunes gift card), and payment schedule (lump sum, per Tweet). Lump sum incentives were paid at the conclusion of the diary; per Tweet incentives were paid at the end of 3-day and 1-week diaries, and at the end of each week of the 4-week diaries. For an overview of the study design, see Table 9.1.

We chose these target populations because they are often hard to reach in surveys but may be more accessible on Twitter. More than one quarter (26%) of 18- to 29-year-old Internet users use Twitter, which is significantly greater than other age groups (14% of 30- to 49-year-olds, 9% of 50- to 64-year-olds, and 4% of those 65 and older). Rates of Twitter use are greatest among the youngest young adults: 31% of 18- to 24-year-old Internet users use Twitter (Smith & Brenner, 2012). We chose Hispanics as the second target population because, at the time the study was designed, Hispanics used Twitter at a greater rate than non-Hispanics. In 2010, 18% of Hispanic Internet users used Twitter, compared with 5% of non-Hispanic white Internet users and 13% of non-Hispanic black Internet users (Smith & Rainie, 2010). Those numbers have since changed, with non-Hispanic blacks now significantly more likely to use Twitter (28% compared with 14% of Hispanics and 12% of non-Hispanic whites) (Smith & Brenner, 2012).The third target population, people with diabetes, is not traditionally identified as a hard-to-reach group, but diabetes is associated with noncompliance to medical interventions such as medications, dietary changes, exercise, as well as with self-report of these, arguing their inclusion as hard to reach.

To help legitimize the study, we included our organization's name, RTI (Research Triangle Institute), in the Twitter handles, used the RTI logo for all of the accounts' profile pictures, and had the official RTI

Twitter account follow each of our seven accounts. Given the complexity of the study and the multiple Twitter accounts, we used a social media dashboard, HootSuite, to manage the Twitter accounts. The dashboard allowed us to toggle seamlessly between accounts, schedule Tweets in advance, and better track communications to and from each Twitter account.

One drawback of using Twitter, however, is the amount of time required to export data after they have been collected. Because of privacy settings of users who did not have public accounts, we could not use any mass extraction tools to export our data. As this study was relatively small, we copied and pasted all of the responses into a spreadsheet. When copying the data, we had to make sure to include the timestamp indicating when the response was received. For a larger Twitter study, exporting all participants' responses by hand could become very time consuming.

Recruitment

We used two online methods for participant recruitment: Twitter and Craigslist. Ads containing a brief description of the study and a link to the screening survey were posted in the volunteers section of Craigslist for major metropolitan areas around the United States. On Twitter, we used the general recruitment account to Tweet links to the screening survey using hashtags[1] relevant to the target populations. We also sent Tweets to key organizations or prominent individuals with large followings of at least one of our target populations. We told these organizations and individuals that we were a nonprofit organization conducting research on Twitter and asked whether they would re-Tweet (RT) the link to our screening survey so their followers would see it. We used a shortened link to conserve space in these Tweets.

The screening survey asked for each participant's Twitter handle (so they could be contacted on Twitter if selected for the study) and demographic information (to see whether they qualified for the study). We used an online survey software tool to create and administer the screening survey. Adults living in the United States who completed the

---

[1] A hashtag (denoted by the # symbol) marks keywords or topics in a Tweet. The # symbol functions as a system for categorizing Tweets by making the marked text more easily searchable on Twitter.

online screening survey, provided their Twitter handle, and fell into one of our three target categories (young adult, Hispanic, people with diabetes) were considered eligible.

Participants were then selected for each diary from completed screening surveys. We carefully selected participants so the final sample was as diverse as possible in demographic characteristics. Some demographics were predetermined depending on the group (young adult, Hispanic), but we also considered other characteristics such as gender, education, employment, marital status, income, and geographic location. We then sent these selected participants a Tweet from the Twitter account created for the applicable study. This Tweet notified them they had been selected to participate in the study and asked them to complete a brief introductory survey by clicking a link in the Tweet.

The introductory survey was not traditional; instead, it provided instructions akin to prenotifications and informed consent. Its main functions were to describe the study, provide instructions on how to participate, and obtain informed consent. (See Figure 9.1 for an example.) Each of the six diaries had its own introductory survey because the diaries differed in topic, incentive, duration, and so on and thus required different descriptions in the informed consent and the instructions. The

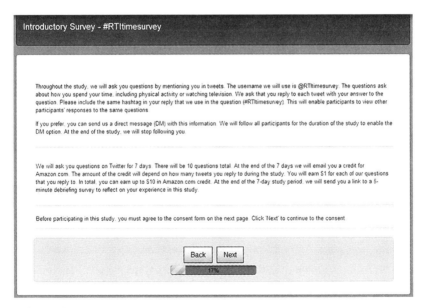

FIGURE 9.1 Introductory survey screenshot.

FIGURE 9.2 Participation process.

introductory survey also asked for the participant's e-mail address so we could send incentives. For diaries that offered a choice of incentives, the introductory survey asked participants to select to receive their payment as an iTunes or Amazon.com electronic gift card. After providing their consent to participate, participants could request to have a copy of the consent form e-mailed to them.

Participants were required to enter their Twitter handle in the introductory survey as their form of identification. After they completed the introductory survey, we asked participants to send a Tweet to the designated Twitter account confirming their participation in the study—this Tweet was the only way we considered their consent to participate to be complete. We required this Tweet because the introductory survey link was visible to others besides the selected participant, and we wanted to be certain that the person who gave consent was the person selected for the study (not someone else entering the Twitter user's handle). If we did not receive a Tweet from the Twitter handle provided in the introductory survey, the individual was excluded from the study.

After the data collection time period for a diary ended, we sent all participants a Tweet with a link to a debriefing survey. We had six different debriefing surveys, one per diary, which gathered feedback on participant experience and was the final portion of the study. A flowchart of the participation process is shown in Figure 9.2.

Data Collection

The diaries asked questions on one of three topics: mood, diet, and activity. Participants assigned to a diet diary were asked to Tweet everything they ate or drank for the duration of the study to which they were assigned. Participants in the mood or activity groups were not asked to record anything on a regular basis, but they were instead Tweeted specific questions—that is, they were mentioned in a Tweet to draw their attention to each question. The questions were asked at seemingly random times throughout the day and night.

Participants in all diaries were given the option to either Tweet their responses (meaning others could view their Tweets) or send them to us via Direct Message (DM). DMs are private messages, still limited to 140 characters, that only the sender and the receiver can see.

We asked participants to use a diary-specific hashtag (e.g., #RTI-foodsurvey) in their Tweets for the study. If someone clicked on one of those hashtags, they were directed to a page where they could see every available Tweet using that particular hashtag. The point of using hashtags in this study was to determine whether participants used them to view responses from other participants in the same diary. We anticipated this ability would make the study more engaging by allowing participants to see how they compared with other participants. We also thought the ability to view others' responses might lead to faster response times (because they could see that other participants were responding). Throughout the study, we sent the participants encouraging messages to keep Tweeting or reminding them to use the hashtag in their responses. At the end of each diary, we sent participants a Tweet thanking them for taking part in the study and asking them to complete the debriefing survey.

Previously, Twitter has been used to help participants track information about themselves,[2] but to the best of our knowledge, no one has conducted diary studies on Twitter to track information about *other* people. We conducted this study to investigate the feasibility of using Twitter in this manner. We varied a number of factors, including topic, incentive, and duration of diary, to gain a general understanding of what will and will not work in a Twitter study. The number of varied conditions coupled with the small sample size limited our ability to compare the options against one another. For instance, we could not conclude that one incentive worked better than another. Rather, we looked for glaring failures of one condition versus another and failures across the study as a whole that would suggest this method of asking questions would be inadvisable in the future.

RESULTS

This section presents the findings of this research: nonresponse rates, the amount of time participants took to respond, the outcome of various

---

[2] Examples include *tweetwhatyoueat* (http://www.tweetwhatyoueat.com/) to track what you eat and weigh, and the now defunct *qwitter*, to track cigarettes smoked and view progress when trying to quit smoking.

TABLE 9.2

Average Number of Question Tweets Replied to and Response Times*

	RTI FoodSurvey ($n = 9$)	RTI Survey2 ($n = 13$)	RTI MoodSurvey ($n = 9$)	RTI TimeSurvey ($n = 12$)	RTI Survey5 ($n = 11$)	RTI Survey6 ($n = 15$)	Total ($n = 69$)
Never responded	0	1	0	1	4	0	6
Mean diet Tweets per day* (Std. Dev.)	3.0 (.7)	—	—	—	1.9 (.6)	—	2.5 (.9)
Mean Qs answered, by participant* (Std. Dev.)	—	84% (25%)	87% (18%)	78% (33%)	—	81% (26%)	82% (30%)
Mean response time (hours) (C.I.)	—	2.9 [2.2,3.9]	2.4 [2.2,2.7]	1.5 [0.9,2.5]	—	2.7 [2.0,3.5]	2.6 [2.2,3.0]

*Excludes participants who never responded.

question formats, incentive preferences, and feedback from participants. We did not observe any obvious failures of our approach of asking questions on Twitter, and in the absence of such obvious failures, we are optimistic that Twitter could work as a platform for future diary studies.

Nonresponse

Unit Nonresponse

A total of 69 participants completed the informed consent, thereby enrolling in the study. As shown in the first row of Table 9.2, we selected six participants for the study, and they provided their consent to participate but never responded as part of their diary. Such unit non-response is interesting, given that we used an opt-in sample and all participants in the study had already put forth some effort in the screening survey and the consent process to participate.

Response Volume

The second and third rows of Table 9.2 display to what extent participants responded, excluding those in row one who never responded. In the diet diaries, response was measured in terms of how many Tweets participants sent us rather than in terms of how many questions they replied to because they were not asked specific questions. Taking into account the different durations of each diary, we calculated the average number of Tweets participants sent per day, as displayed in row two. Participants in RTIFoodSurvey and RTISurvey5 (the two diet diaries) sent an average of 3.0 and 1.9 Tweets per day, respectively. Interestingly, RTIFoodSurvey lasted more than twice as long as RTISurvey5 (7 days versus 3 days), but participants sent more Tweets per day. Although these numbers are indicators of the participants' accessibility, they do not necessarily reflect the amount of information received because some participants reported multiple meals in a single Tweet.

The third row displays the percentage of questions participants responded to, on average, by diary. Excluding the diet diaries, participants responded to 82% of the questions they were asked on Twitter. The rate differed by diary and ranged from 78% to 87%.

Response Times

The fourth row of Table 9.2 shows, for the questions that participants answered, the average number of hours they took to respond. These data, like most time data, are skewed, with most people taking a short time and a few people taking a long time. Therefore, we calculated the means using log transformations to adjust for the skewed distribution of the data. Overall, participants answered the questions an average of 2.6 hours after they were asked, with differences across diaries ranging from 1.5 hours for RTITimeSurvey to 2.9 hours for RTISurvey2.

We reviewed the average response times for each question to see how they varied by time the question was asked.[3] In Figure 9.3, each point on the scatterplot represents one question asked in the specified diary. The shortest response times tended to occur in response to questions that were asked between 8:00 AM and 1:00 PM.

We examined the average response times by order of the question in the diary (Figure 9.4). Response times fluctuated across questions within each diary. Some of the peaks were expected because the questions

[3] The time a question was asked refers to the participants' time zone, not necessarily to our own time zone. We sent the questions on a rolling schedule so each participant in a study received the question at the same time, regardless of the time zone in which he or she lived.

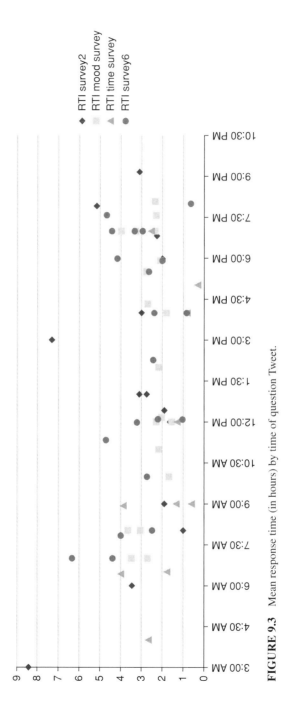

FIGURE 9.3 Mean response time (in hours) by time of question Tweet.

214

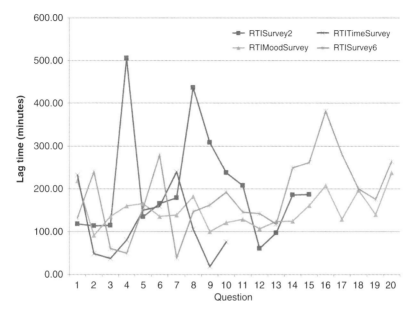

FIGURE 9.4 Mean response time (in hours) by question.

were asked at inconvenient times (e.g., 3:00 AM for Question 4 on RTI Survey2), whereas others were unexpected (e.g., 3:00 PM for Question 8 on RTISurvey2).

Response Times by Age

Table 9.3 shows average response times by age. On average, 18- to 40-year-olds responded in less than 2 hours, whereas 41- to 50-year-olds responded in 2.7 hours and 51- to 60-year-olds in 3.4 hours. We did not test for significance because of the low cell counts and because age is

TABLE 9.3
Average Response Time in Hours, by Age

	18–24 ($n = 27$)	25–30 ($n = 7$)	31–40 ($n = 7$)	41–50 ($n = 4$)	51–60 ($n = 2$)
Response time	2.0	1.3	1.9	2.7	3.4
Confidence interval	[1.2, 3.1]	[0.6, 2.7]	[0.8, 3.8]	[0.8, 9.9]	*

*Sample size is too small to report the confidence interval.

confounded with the topic, duration, and incentive to which participants were assigned.

Response Rates by Question Format

The activity and mood diaries asked questions in a variety of formats to gain insight into how well or poorly the different formats worked with Twitter. Table 9.4 shows the types of question formats and an example of each, followed by the percentage of questions in each format that were answered, by diary. The response rates varied by diary, but given the study design, it is unclear how much of the variation (if any) is attributable to the format.

Data Quality

We asked questions using a variety of question formats to see how they would work on Twitter. Overall, the responses aligned with our expectations and would be suitable for analysis. However, for this study, we focused our analysis on the quality of responses rather than on analyzing the information obtained through the responses themselves. Looking at the quality of responses participants gave to each type of question, we conclude that none led to major problems, but some worked better than others. This section discusses the quality of open-ended and closed questions.[4]

Open-Ended Questions

The only problem we observed with open-ended questions was that on occasion, when we asked two questions in the same Tweet, participants answered only one of them. We recommend omitting this format from future Twitter surveys and limiting questions to one per Tweet. Otherwise, open-ended questions worked particularly well on Twitter, perhaps because communication on Twitter is free-flowing, so open-ended responses looked similar to the regular (nondiary) Tweets participants may send. We observed that many participants gave us more information than we asked for in their responses to open-ended questions. For

---

[4] See Table 9.4 for examples of open-ended (rows A and B) and closed (rows C–H) questions.

TABLE 9.4

Response Rates by Question Format

Question Format	Example	RTI Survey 2 (n = 12)	RTI Mood Survey (n = 9)	RTI Time Survey (n = 11)	RTI Survey 6 (n = 15)	Total (n = 49)
Open-ended						
A. 1 question	What were you doing from 3:00PM-4:00PM today?	78% (7)	89% (4)	75% (6)	85% (4)	80% (21)
B. 2 questions	The last time your heart was beating fast or you got out of breath, what were you doing? When was it?	92% (2)	87% (6)	73% (1)	79% (6)	82% (15)
Time						
C. ___ hours or minutes	Yesterday, how many hours or minutes did you spend watching TV?	91% (3)	—	82% (3)	—	87% (6)
D. ___ hours & ___ minutes	How much time did you spend watching TV yesterday? ___ hours & ___ minutes.	67% (1)	—	—	—	67% (1)
E. Hours with 15 minute increments as fractions of an hour	Yesterday, how many hours did you spend sitting? (15 min = 0.25 hrs., 30 min = 0.5 hrs., 45 min = 0.75 hrs.)	79% (2)	—	—	—	79% (2)

(Continued)

217

TABLE 9.4
(*Continued*)

Question Format	Example	RTI Survey 2 (n = 12)	RTI Mood Survey (n = 9)	RTI Time Survey (n = 11)	RTI Survey 6 (n = 15)	Total (n = 49)
Select one response option						
F. Select one	How is your mood today? Very good, somewhat good, neither good nor bad, somewhat bad, or very bad?	—	80% (3)	—	78% (3)	79% (6)
Scale						
G. 0 to 100	If 0 = extremely sad, 50 = neutral, and 100 = extremely happy, how sad or happy do you feel right now?	—	82% (5)	—	81% (5)	82% (10)
H. 1 to 7	On a scale of 1 to 7, how were your spirits today? 1 = extremely poor, 7 = extremely good.	—	89% (2)	—	87% (2)	88% (4)
Total		84% (15)	87% (20)	78% (10)	81% (20)	82% (65)

*n represents the number of participants and excludes those who never responded.
**The number of questions asked is displayed in parentheses.

218

example, we received the following responses[5,6] to the question, "What did you do yesterday?"

> *Went to the new aquarium in [CITY],* **had a fantastic day with family**

> *I slept in, cleaned, spent time outdoors, played w my kids and cats, had a picnic in the park!* **Fun times.**

> *Yesterday I watched football, prepped for Sunday School, took a nap.* **I didn't get to work or the gym.**

> *Went shopping* **to buy stuff for my trip.** *Had brunch w/ my bf and his family* **for his b-day.**

This extra information that participants provided may be useful in some studies. Even if the researchers do not need supplemental information from participants, providing this information may keep participants more engaged in the study and less likely to drop out. The following example from one of the diet diaries demonstrates how one participant became more engaged in the study through his own commentary:

> *Tonite went Mediterranean with chicken kebab and tabouli.* **The homemade hummus was to die for!**

Even though many participants added extra information to their open-ended responses, their responses tended to be concise. This observation was not surprising given the character limit of Tweets. Participants in the diet diaries were especially concise; some participants reported multiple meals in a single Tweet, as illustrated in the following responses:

> *Breakfast: Water & muffin. Lunch: Ham sandwich, Doritos, cookies, banana and water. Now eating crackers and water.*

> *Breakfast: Pastry with Milk – Lunch: Pasta and Milk/Tea Dinner: Chicken Fingers, Potatoes, and Grape Juice*

Although the character limit could prevent participants from giving a complete response, we found they were able to answer our questions

---

[5] Bold font was added to emphasize the additional information participants included in their responses.

[6] To protect the identities of participants, we revised all Tweets in this chapter to change the content without altering the meaning.

fully in a single Tweet. Also, participants added extra information only when they had already answered the question and still had room for additional characters. For instance:

Had a late night snack: a bowl of Cheerios and granola. **I hate when I'm so busy I can't sit down for normal meals***.*

For open-ended questions that require a more detailed and lengthy response, the character limit may be restrictive. However, for most questions, the character limit is actually advantageous because it forces participants to answer the questions directly and think through their responses before expressing them.

Closed Questions

Compared with open-ended questions, closed questions elicit responses that are much less Tweet-like and, to a certain extent, unnatural on Twitter. Even though these sorts of responses are somewhat incongruous on Twitter, the quality of responses to our closed questions was adequate and we observed no major problems relating to data quality.

Typically, a benefit of using electronic methods to collect self-reported data is that the instrument can be programmed to require responses to be in the desired format. For instance, when participants are prompted to provide a number, the instrument can be programmed to accept only whole numbers falling within a certain range. These restrictions are not possible on Twitter, so the response data to closed questions would likely need to be cleaned before analysis. Below are examples of the different responses we received to the question, "On a scale of 1 to 7, how were your spirits today? 1 = extremely poor, 7 = extremely good." Most participants provided the desired response (a whole number between 1 and 7), but several added punctuation and/or commentary, and one responded with a decimal number:

5.5

definitely a 4

1!!!

I feel great! 7

About a 5. It was a day of relaxation.

I was at ~6 today. Would've been at 7 if it hadn't rained!

Probably 5. Was a long day, but I'm going out with friends soon. Really excited for it.

Most of these responses could easily be cleaned for analysis by extracting the number each participant provided. Depending on how the analysis is done, however, the 5.5 could be difficult. In this instance we could have asked the participant if his or her answer was closer to 5 or 6, much as an interviewer would do in a standardized survey interview. Such follow-up would prevent us from having to drop the response, which might be necessary depending on how the analysis is done. Fortunately, even though Twitter allows free-form responses to closed questions, replying to the Tweet to request clarification is easy. However, this process could be time-consuming and participants may not respond.

Conversational Responses

As several of the Tweets discussed previously demonstrate, many responses were conversational and included additional, unnecessary information for both open-ended and closed questions. For instance, one participant started his Tweet with "Good morning!" before listing what he had for breakfast. Following up with the most conversational participants to get more information, including clarification on previous responses, would be fairly easy. Twitter could be a great format for studies that need back-and-forth communication between researchers and participants.

Incentive Preference

Some diaries provided a guaranteed incentive for participants, and we asked participants in those diary studies whether they preferred an Amazon.com gift card or an iTunes gift card. Participants who received the incentive per Tweet were given an Amazon.com gift card. For those participants iTunes was not an option because its gift cards are only available in certain increments, while Amazon.com gift cards are available in any amount. Of those given a choice of incentive, all but one participant selected the Amazon.com option. We assume participants overwhelmingly preferred Amazon.com because of its greater selection

TABLE 9.5

Participants Who Completed Debriefing Survey by Diary

RTI FoodSurvey ($n = 9$)	RTI Survey2 ($n = 13$)	RTI MoodSurvey ($n = 9$)	RTI TimeSurvey ($n = 12$)	RTI Survey5 ($n = 11$)	RTI Survey6 ($n = 15$)	Total ($n = 69$)
100%	69%	100%	100%	91%	93%	88%

of items beyond music, videos, and apps (the main items in the iTunes store).

Participant Feedback

The debriefing survey we gave to participants at the end of the study provides perhaps the most interesting insight into the utility and potential benefits of asking questions on Twitter. A total of 88% of participants selected for the diary studies completed the debriefing survey, including—surprisingly—three of the four nonrespondents in RTISurvey5. The percentage of participants who completed the debriefing survey by diary is displayed in Table 9.5.

Nonresponse

In the debriefing survey, we asked participants who did not reply to all of the questions why they did not reply. They reported that they were not notified by Twitter that we had mentioned them in a Tweet. This lack of notification could result from participants' personal Twitter settings, the dashboard or app they were using to receive Tweets, or issues directly with Twitter. We asked participants in the diet surveys a similar question. One participant in RTISurvey5 who never responded said it was because she "did not think to Tweet these things."

Response Times

When we asked how often they replied immediately after seeing a question Tweeted at them, 19% of participants said *always*, 50% said *usually*, 29% said *sometimes*, and 2% said *rarely*. Participants who did not always reply immediately were asked why. Most said they were busy or at work when they first saw the question, an understandable explanation. One participant in RTIsurvey6 (topic: mood), however, gave a particularly troubling explanation in regard to potential error. The participant's explanation was "wanting my day to go in a certain way before I responded." This participant responded to 95% of the questions asked and answered them all by Tweet rather than DM. She might have taken the same approach in a typical (non-Twitter) diary study, but her tendency to wait until she could respond in a certain way may have been exacerbated by the public nature of Twitter and the her preference to Tweet instead of DM.

Participants in the diet diaries were asked how often they Tweeted immediately after eating or drinking something. A total of 11% said *always*, 50% said *usually*, 39% said *sometimes*, and no one said *rarely* or *never*. Participants who did not always Tweet immediately after eating or drinking had various explanations, including the main reasons given for the nondiet diaries: They were busy or at work. Diet diary participants also reported they did not Tweet immediately after eating or drinking because they did not have Internet access, their phone was not working, or they forgot. One participant who did not always Tweet immediately reported keeping records of information to Tweet later. The participant explained, "My work and school schedule prevented me from Tweeting immediately afterwards. I didn't always have access to Twitter, but I kept a small notebook with me to write down what I ate and drank during these times."

Public Nature of Twitter

The public nature of Twitter changes the traditional assumptions of survey interviewing and makes it a fundamentally different environment. Completing a survey is normally a private process; even when interviewers are required to collect the data, they take extensive steps to ensure the confidentiality and the privacy of participants' responses. Such assurance is not possible with Tweets, which are visible to the user's followers or the general public, depending on individual privacy settings. Participants can DM their responses to make them visible only to the survey organization's account, but sending a DM requires an extra step beyond just responding to a Tweet. With survey response taking place in such a public sphere, results may be biased by social desirability effects. Some participants may answer differently on Twitter than they would in a different mode.

A social desirability effect might dampen response to sensitive or quasi-sensitive items (such as mood, activities, or food intake) by reducing reporting of embarrassing feelings or unhealthy meals, for example. On the other hand, the public aspect of Twitter could encourage response by exerting social pressure and motivating participants. For instance, Twitter participants might feel more accountable for reporting their answers knowing (and observing) that other users in the study are using the hashtag and responding via Twitter.

In our exploratory study, although most participants (56%) said the answers they reported were *not at all* influenced by the possibility that other people would see their Tweets, many participants still expressed concerns. A total of 25% said their responses were influenced by the public nature of the study *only a little*, 13% said *some*, and 7% said *a great deal*. This concern was also evident in a later debriefing question that asked what participants disliked about the study, if anything. Most participants said they disliked "nothing," but five participants stated that they did not want their responses to be public. Interestingly, four of these five participants were part of RTISurvey6, the 4-week mood study of young adults.

Tweeting Versus Direct Messaging

Because all participants were given the option to DM their responses, we were surprised that five participants said they disliked the public aspect of the study. Four of these five participants Tweeted all of their responses. We assume they may not have read the instructions carefully when they were accepted into the study and assumed they were required to Tweet instead of DM. Or perhaps they did not know how to DM because this function is underused on Twitter. This problem was not widespread, however, as 19% of participants responded entirely via DM and 11% responded with a combination of both Tweets and DMs. Most participants, 70%, Tweeted all of their responses.

Participants who responded only by DM were asked why; they replied mainly that they did not want to clog their friends' Twitter feeds with extraneous Tweets, but some cited privacy concerns. Also, one participant in RTISurvey5, the diet survey of young adults, explained, "I'm not particularly proud of my diet! Nor do I think most people want to know."

Others in the diet diaries may have shared that participant's sentiment. As shown in Table 9.6, a marginally significant difference was

TABLE 9.6
Response Mode by Diary Topic†

	Activity ($n = 23$)	Mood ($n = 24$)	Diet ($n = 16$)
Only Tweets	78%	79%	44%
Only DMs	13%	8%	44%
Both Tweets and DMs	9%	13%	13%
Total	100%	100%	100%

† $\chi^2(4, N = 63) = 9.29, p = 0.054$.

evident in mode of response by diary topic, $\chi^2(4, N = 63) = 9.3$, $p < 0.10$. Participants in the diet diaries were less likely to Tweet their responses than participants in the activity and mood diaries. Furthermore, we noticed an interesting contrast in responses that were Tweeted compared with those sent as a DM. The responses in Figures 9.5 and 9.6 are from a participant with diabetes who tended to Tweet the healthy food he ate and DM the unhealthy food.

JohnJohn7923 @JohnJohn7923
@RTIFoodSurvey apple, yogurt, glass of water #rtifoodsurvey
4:06pm, 25 Feb 12

JohnJohn7923 @JohnJohn7923
@RTIFoodSurvey applesauce #rtifoodsurvey
2:20am, 25 Feb 12

JohnJohn7923 @JohnJohn7923
@RTIFoodSurvey diet coke #rtifoodsurvey
4:43pm, 24 Feb 12

JohnJohn7923 @JohnJohn7923
@RTIFoodSurvey a protein bar and a bottle of water
12:25pm, 23 Feb 12

JohnJohn7923 @JohnJohn7923
@RTIFoodSurvey a fruit cup and a glass of OJ #rtifoodsurvey
11:40am, 20 Feb 12

FIGURE 9.5   Example Tweets diary entries.[7]

---

[7]The image and username are fictitious, used only for illustrative purposes. To protect the participant's identity, we revised the Tweets to change the content without altering the meaning.

JohnJohn7923 @JohnJohn7923 🐦
diet dr. pepper doritos and some pop corn
10:50pm, 24 Feb 12

JohnJohn7923 @JohnJohn7923 🐦
3 pieces of pizza #rtifoodsurvey
3:18pm, 24 Feb 12

JohnJohn7923 @JohnJohn7923 🐦
hot dog, mac n cheese, carrots, and diet dr. pepper
6:34pm, 23 Feb 12

JohnJohn7923 @JohnJohn7923 🐦
5 oreos, orange juice, peanut butter toast
1:26am, 21 Feb 12

JohnJohn7923 @JohnJohn7923 🐦
Went to McDonald's and got a crispy chicken sandwich, a dr. pepper, and fries.
10:30pm, 20 Feb 12

FIGURE 9.6   Example DMs.

General Feedback

Aside from some participants' reservations about participating publicly, participants liked the study. When asked in an open-ended question what they liked most, they used three main descriptors: easy, simple, and convenient. The diaries used in this study were probably more basic than in typical diary studies, and we wonder whether participants would use the same descriptors if participating in a more extensive Twitter diary.

Hashtags

We asked participants to use diary-specific hashtags in their responses, thinking that if they clicked on or searched for the hashtags they would access other participants' activity, which in turn might keep them more engaged in the study and participating more regularly. We were unable to track whether participants clicked on or searched for the hashtags, so we used self-reported data of hashtag use instead. In the debriefing survey, fewer than half of participants (39%) said they used the hashtags to see how other participants responded.

DISCUSSION

Twitter offers a free and readily accessible platform for conducting diary studies. Learning how to use Twitter is easy, so study participants do not need to be limited to current Twitter users. Twitter is accessible from participants' own Internet-capable devices, so participants may update their Twitter diaries more regularly, unlike diaries on provided devices that are more easily forgotten. In this exploratory study, we did not compare response times on Twitter with response times on other diary modes, but we were impressed with how quickly participants responded to our questions on Twitter.

Twitter diaries could work effectively for nearly anyone who knows how to use a computer or smartphone, but the diaries are especially suited for collecting data from people who are active Twitter users because Tweeting is already routine for them. We envision Twitter used as a mode option for multimode surveys. Sample selection could be done using traditional survey sampling techniques. In some studies, recruiting a convenience sample on Twitter would be acceptable, especially when one is interested only in studying people on Twitter, or when conducting health interventions.

The main benefit we found was participants' accessibility to us as researchers. On average, participants answered our questions within 2.6 hours, a remarkably low response time considering questions were asked at all hours of the day and night. We attribute the quick response times to Twitter's accessibility across many devices and to the participants' high use of these devices (91% Tweet from a computer, 86% from a smartphone, and 28% from a tablet).

The following list provides reminders and recommendations for future Twitter diaries, based on our findings from this exploratory study:

1. Although a social media dashboard simplifies the process of managing Twitter accounts, setting up all the Tweets and downloading responses is still time consuming.
2. Twitter's main benefit is that participants become accessible to researchers; however, participants expect the study administrator to be accessible at all times too. Thus, the study administrator or researcher must monitor the Twitter accounts every day, including weekends.

3. In this study we sent the questions as Tweets and asked participants to Tweet or DM in return. If we conducted such a study again, we would send the questions as DMs and ask participants to reply via DM. DM replies might alleviate privacy concerns without sacrificing Twitter's quick access to participants. Furthermore, we suspect this tactic may make our Tweets more noticeable while allowing participants to track easily which Tweets they have answered because most people Tweet much more than they DM. Consequently, their DM feed is less full, and tracking incoming and outgoing messages is easier. Note: A person must be following you on Twitter before you can send them a DM, so the initial contact must take place via Tweets or some other means.

4. Comparing the effectiveness of the different incentive amounts and payment structures is difficult because of the number of variables in this study. To simplify the provision of incentives, we recommend offering a guaranteed amount instead of an amount per Tweet. Because Amazon.com was much more popular than iTunes, we also recommend Amazon.com credits only, eliminating the choice between two incentives. Overall, each incentive structure worked reasonably well and future studies on Twitter should examine their impact more closely.

5. Future diary studies should not use hashtags to mark relevant Tweets. Participants often forgot to use them in their replies, and most said they did not use the hashtags to see other Tweets from the study. Furthermore, hashtags took up valuable space in the 140-character limit. However, hashtags may be more useful in studies that collect data or engage participants for another purpose, for example, as part of a support group on Twitter.

This exploratory study was designed to use a variety of approaches to conducting diary studies on Twitter. The study design does not enable direct comparisons of one approach versus another to conclude which is better. Many unanswered questions remain about conducting diary studies on Twitter, including:

- How many questions can we ask in a day without experiencing a decline in response rates?

- For how long are people willing to participate on Twitter?
- What incentive structure maximizes response rates?
- To what extent are responses on Twitter affected by social desirability bias?
- What utility comes from examining Twitter paradata, such as social connections and Tweets from before and after the diary period?

Although we cannot answer specific questions about which approach is best, the study design is helpful because it gives an overview of whether this approach of collecting diary data on Twitter is both feasible and beneficial. Based on the findings from this exploration, we think Twitter is an avenue worth exploring for certain types of studies.

REFERENCES

ALFVEN, G. (2010). SMS pain diary: A method for real-time data capture of recurrent pain in childhood. *Acta Paediatrica, 99*(7), 1047–1053.

ANHOJ, J., & MOLDRUP, C. (2004). Feasibility of collecting diary data from asthma patients through mobile phones and SMS (short message service): Response rate analysis and focus group evaluation from a pilot study. *Journal of Medical Internet, Research, 6*(4), e42.

BEVILLE, H. M. JR., (1949). Surveying radio listeners by use of a probability sample. *Journal of Marketing, 14*(3), 373–378.

BOLGER, N., DAVIS, A., & RAFAELI, E. (2003). Diary methods: Capturing life as it is lived. *Annual Review of Psychology, 54*, 579–616.

FOX, J. F. (1934). Leisure-time social backgrounds in a suburban community. *Journal of Educational Sociology, 7*, 493–503.

HAWORTH, J. (2010). The way we are now. *Leisure Studies 29*(1), 101–110.

JAMISON, R. N., RAYMOND, S. A., LEVINE, J., SLAWSBY, E. A., NEDELIJKOVIC, S. S., & KATZ, N. P. (2001). Electronic diaries for monitoring chronic pain: 1-year validation study. *Pain, 91*(3), 277–285.

MARWICK, A. E., & BOYD, D. (2010). I Tweet honestly, I Tweet passionately: Twitter users, context collapse, and the imagined audience. *New Media & Society, 13*(1), 114–133.

MUNDT, J. C., PERRINE, M. W., SEARLES, J. S., & WALTER, D. (1995). An application of interactive voice response (IVR) technology to longitudinal studies of daily behavior. *Behavior Research Methods, Instruments & Computers, 27*(3), 351–357.

PHILLIPS, M. L., HALL, T. A., ESMEN, N. A., LYNCH, R., & JOHNSON, D. L. (2001). Use of global positioning system technology to track subject's location during environmental exposure sampling. *Journal of Exposure Analysis and Environmental Epidemiology, 11*(3), 207–215.

Semiocast (2012). *Twitter reaches half a billion accounts: More than 140 millions [sic] in the U.S.* Retrieved from http://semiocast.com/publications/2012_07_30_Twitter_reaches_half_a_billion_accounts_140m_in_the_US.

SEVICK, M. A., STONE, R. A., ZICKMUND, S., WANG, Y., KORYTKOWSKI, M., & BURKE, L. E. (2010). Factors associated with probability of personal digital assistant-based dietary self-monitoring in those with type 2 diabetes. *Journal of Behavioral Medicine, 33*(4), 315–325.

SHRIER, L. A., SHIH, M. C., & BEARDSLEE, W. R. (2005). Affect and sexual behavior in adolescents: A review of the literature and comparison of momentary sampling with diary and retrospective self-report methods of measurement. *Pediatrics 115*(5), e573–e581.

SMITH, A., & BRENNER, J. (2012). *Twitter use 2012.* Retrieved from http://pewinternet.org/Reports/2012/Twitter-Use-2012.aspx.

SMITH, A., & RAINIE, L. (2010). *8% of online Americans use Twitter.* Retrieved from http://www.pewinternet.org/Reports/2010/Twitter-Update-2010/Findings.aspx.

SUDMAN, S. (1964). On the accuracy of recording of consumer panels: Part I. *Journal of Marketing Research, 1*(2), 14–20.

TAYLOR, C. (2012). *Does Twitter have half a billion users?* Retrieved from http://mashable.com/2012/07/30/twitter-users-500-million/.

Twitter (2012a). *What is Twitter?* Retrieved from https://business.twitter.com/en/basics/what-is-twitter/.

Twitter (2012b). *Twitter turns six.* Available at http://blog.twitter.com/2012/03/twitter-turns-six.html.

WIEHE, S. E., CARROLL, A. E., LIU, G. C., HABERKORN, K. L., HOCH, S. C., WILSON, J. S., & FORTENBERRY, J. D. (2008). Using GPS-enabled cell phones to track the travel patterns of adolescents. *International Journal of Health Geographics, 7*, 22.

WILLIAMS, P., & STRAUS, R. (1950). Drinking patterns of Italians in New Haven: Utilization of the personal diary as a research technique. I. Introduction and diaries 1 and 2. *Quarterly Journal of Studies on Alcohol, 11*, 51–91.

CHAPTER 10

Recruiting Participants with Chronic Conditions in Second Life

Saira N. Haque and Jodi Swicegood, *RTI International*

Second Life (www.secondlife.com) is a virtual world where individuals represent themselves through avatars. (An avatar is a graphical representation of the user that can exist as an extension of self or as an alternate character.) Once created, the physical appearance of an avatar can be customized, and some users choose to create multiple avatars. The purposes of Second Life include socializing, entertainment, education, and role-playing. As of March 2013, Second Life had more than 29 million registered accounts.

Second Life has promise as a means of studying patients with chronic medical conditions. Unlike acute illnesses, people with chronic conditions live with symptoms for years. In addition, the severity of many chronic conditions has the potential to render part of the population homebound. In a study titled *Recruiting Special and Hard-to-Reach Populations in Second Life*, we discovered that communities have arisen organically in Second Life to support users with these conditions (Swicegood et al., 2012). Through their avatars, users experience presence—as

Social Media, Sociality, and Survey Research, First Edition.
Edited by Craig A. Hill, Elizabeth Dean, and Joe Murphy.
© 2014 John Wiley & Sons, Inc. Published 2014 by John Wiley & Sons, Inc.

they focus on their avatar, the way it moves by walking, running, or flying, and its interactions with others—users experience varying levels of immersion in their virtual surroundings. As online communities allow for the exchange of experiential and medical information, many provide long-term support to patients with chronic conditions in Second Life (Gorini et al., 2008). Although disease-specific online resources such as message boards and microblogs exist in many web-based platforms, they do not allow for integration with other aspects of online life in the way Second Life does because they may only focus on singular topics. Users with chronic conditions face multiple challenges in the healing process. The complexities of these conditions can be explored in a singular location, such as the virtual world Second Life.

We sought to learn more about Second Life *residents*, or users, with chronic medical conditions such as chronic pain, diabetes, HIV+ status, and cancer to support the design and implementation of targeted health interventions. The purpose of this research was to determine the feasibility of recruiting these populations in a virtual world setting such as Second Life. This research also helps us understand the demographics of those with these conditions in Second Life. In turn, these data afford a better understanding of how these populations can be effectively recruited. We chose the four chronic conditions because of their wide range of severity, in addition to changes in lifestyle known to impact the disease process, such as impaired mobility. We were also unsure whether lifestyle changes, such as those caused by limited mobility, would affect the likelihood of virtual world use. With future health interventions for U.S. residents in mind, we also chose these conditions because of their high rate of prevalence among the general U.S. population.

Some survey work has been conducted in Second Life, but it has not focused on individuals with chronic conditions. Determining which methods are successful in recruiting populations with chronic conditions helps us understand how to target them in Second Life. In turn, this knowledge can be used to facilitate the design and implementation of targeted interventions in Second Life with real-life health implications. Thus, interventions in Second Life can potentially influence healthy behaviors such as diet, exercise, and medication adherence to impact health outcomes in a positive way (Siddiqi et al., 2011).

Second Life has shown promise for managing chronic conditions (Novak, 2010); yet more information is needed to understand how to recruit members of these populations. We used methods that targeted our chronic-condition populations (in addition to more general methods). This study was conducted in both a real-life setting—a web-based setting outside Second Life—and from within the virtual environment. Previous studies conducted in this setting often relied on convenience samples or strictly real-world recruiting, not necessarily on broad-based recruiting within Second Life. This chapter outlines the recruitment methods used and their success. Effective methods can be used to identify best practices for recruiting in Second Life for both general and chronic-condition populations in Second Life, in addition to providing recommendations for others conducting similar studies.

BACKGROUND

Recruitment methods that have been successfully implemented in other settings may not work in Second Life because of the environmental and contextual differences of conducting research in a virtual world. We were interested in which recruitment methods were most effective for those with chronic conditions.

Previous studies in Second Life have used various recruitment methods including Second Life classified ads and onsite or inworld recruitment (Bell, 2009; Bell et al., 2009; Chesney et al., 2009; Dean et al., 2011; Foster, 2011; Novak, 2010; Siddiqi et al., 2011). One study recruited participants through a Second Life classified ad that provided a teleport—or instantaneous transportation—to a signup kiosk. Once the link in the ad was initiated, or clicked, a map of the location of the signup kiosk was shown. At this point, users could accept or decline a teleport to an alternate location inworld. The ads ran for 10 months and yielded 5,600 teleports, of which 5,265 signed up to participate in research studies. Overall, the cost-per-teleport was reported as L$38 (Linden dollars, the Second Life currency) or approximately US$0.15 (Novak, 2010).

Other methods of inworld recruitment have included postings to the Second Life Forum—a community discussion board—and advertising on a billboard in a virtual shopping center. Inworld newspapers have also been used to recruit study participants (Dean et al., 2011). Collectively these methods yielded 114 Second Life users or 34.5% of total participants (Dean et al., 2011). Instant inworld messaging has been used to target members of selected groups in Second Life, and web e-mail campaigns have proven helpful for targeting Second Life users with related interests (Bell, 2009; Dean et al., 2011).

Also used in virtual-world research, onsite recruitment typically includes searching for heavily populated locations inworld and approaching avatars individually or as a group. The techniques used in virtual health studies have included contacting health-related interest groups in Second Life to assist with advertising for a study in the Second Life Community Events calendar and at related conferences and locations inworld. Participant referrals have been used to recruit virtual-world participants as well as messaging members of an intervention-related group in Second Life (Siddiqi et al., 2011). Real-life methods have included posting recruitment materials in newspapers, e-mailing topic-related listservs, and posting messages in Second Life blogs (Siddiqi et al., 2011).

Siddiqi, Mama, and Lee recruited a total number of 162 participants out of 239 avatars approached within a 6-month recruitment period through onsite recruitment. This method yielded a relatively low recruitment ratio of 19.7%. These findings also reported that most participants signed up for the study without being approached by a recruiter and that, on average, 1.3 hours of time was required to enroll one participant (Siddiqi et al., 2011).

Our study used previously explored methods, in addition to other experimental techniques, to recruit chronic-condition populations in Second Life.

METHODS

We developed a recruitment protocol based on best practices in recruiting special and hard-to-reach populations as well as methods that have previously been used in virtual-world research. We modified the protocol as necessary during data collection to include more targeted approaches to recruit members of our four chronic-condition populations. Second Life users in the populations of interest were recruited to complete a health survey. Data were collected through an online web-based survey with questions about respondents' technology use, exercise habits, specific medical conditions, and overall health and wellness. After Institutional Review Board (IRB) approval was secured, data collection took place between December 2011 and April 2012.

Instrument Development

We used two instruments in this study. First, an eligibility survey was administered to determine whether individuals met the criteria for inclusion: that all participants were at least 18 years of age, U.S. residents, and currently had or had ever had one or more chronic conditions of interest in real life. If individuals met the criteria, we administered a health-related survey. Both were administered online and accessed through a participant's private web browser.

We were interested in how people used the Internet, social media, and other technologies and whether that use differed across populations. Because our questions were not related to virtual worlds, we focused on general technology use as a proxy for willingness to use technology for health purposes. The survey was developed with questions from technology-related surveys conducted by the Pew Research Center, as well as four health-related surveys: The National Survey of Drug Use and Health (NSDUH), The National Health Interview Survey (NHIS), The National Survey of Family Growth (NSFG), and The National Health and Nutrition Examination Survey (NHANES). During questionnaire development, the Pew surveys were searched for items related to the use of mobile devices to access the Internet. The health surveys were searched for items relating to health conditions (diabetes, chronic pain, HIV/AIDS, cancer, and overall health and wellness) and participation in support groups. All relevant items found on these topics were considered for inclusion in the survey. A team of five reviewers narrowed down the list of selected items to those that fit within the overall purpose of recruiting chronic-condition populations in Second Life.

Recruitment Methods

Because previous findings reported the time commitment and low recruitment ratio of direct or onsite recruitment, this study relied on other methods (Foster, 2011; Siddiqi et al., 2011) including a Second Life blog, Second Life classifieds, and a Second Life Forum discussion board. Word-of-mouth and other participant-generated referrals were also incentivized through a referral program. Targeted recruitment was conducted by contacting several health-related support communities that have formed to support chronic-condition populations inworld. Through these groups, recruitment messages were posted at health-related

TABLE 10.1
Recruitment Type and Setting

	Recruitment Type	
Setting	General	Targeted
Real Life	Craigslist.com, Facebook & Twitter, Second Life Blog	N/A
Second Life	Second Life Classifieds, Second Life Forum	Referral program, information sessions, general networking

locations and mass-messaged to group members. Recruitment-specific information sessions were advertised in the Second Life Community Events calendar and conducted at two inworld support communities.

Although virtual-world participants have been successfully recruited for health-related studies in Second Life (Foster, 2011; Murphy et al., 2010; Siddiqi et al., 2011; Swicegood et al., 2012), this study explored additional techniques including Facebook and Twitter. We used other forms of social media including a Second-Life–related blog and its established connections on Facebook, Twitter, Plurk, and Google+. Other real-world recruitment efforts included placing classified ads in major metropolitan cities on Craigslist.com.

As shown in Table 10.1, general real-life recruitment methods included classified ads placed on Craigslist.com and posting recruitment messages on Facebook, Twitter, and a Second-Life–oriented blog. General recruitment methods administered in Second Life included Second Life classifieds and a post on the Second Life Forum. Targeted recruitment was applied within Second Life. These methods include a referral program, recruitment-specific information sessions, and general networking among health-related support communities and their leaders in Second Life.

All recruitment materials contained the name of the recruitment coordinator's avatar, in addition to the project director's real name and phone number. Although contact information is typically required as part of informed consent procedures, researchers also felt it necessary to include Second Life contact information for users who might be hesitant about revealing their real identity or contact information to the project team.

General Recruitment

General recruitment allowed those with chronic conditions of interest who do not participate in activities related to their disease in Second Life the opportunity to participate. Several of these methods have been previously used in Second Life research including ads on Craigslist.com, Second Life classifieds, the Second Life Forum, and word-of-mouth recruiting including participant-generated referrals (Bell, 2009; Dean et al., 2011; Foster, 2011; Head et al., 2012; Richards & Dean, 2012).

We used Craigslist.com to post weekly recruitment messages. Initially, ads were placed in Craigslist-specified major metropolitan cities including Boston, Chicago, Los Angeles, New York City, Portland, Sacramento, San Diego, Seattle, San Francisco, and Washington, DC. These ads expired after 7 days. After the list of major metropolitan cities was exhausted, other major cities with Craigslist were targeted and recruitment messages were posted in alphabetical order—as listed on the Craigslist website. These included Atlanta, Austin, Dallas, Denver, Detroit, Houston, and Las Vegas. Because Craigslist restricts the placement of similar ads in multiple cities, ads were posted individually and reposted weekly. In addition, this method required the development of recruitment messages, but it was not costly or time consuming.

Weekly inworld classified ads were also used to recruit participants. Second Life classifieds were posted in the "Employment" and "Personal & Wanted" sections in Second Life. Second Life users could search for specific keywords, such as "survey" or "research" to view our ad. Figure 10.1 demonstrates how Second Life users could locate our ad among other search results. When searching "survey," our ad is ranked first.

All ads are cost sensitive—search results are listed according to how much users pay for posting—and information regarding the cost of similar ads is visible when posting. These data allowed us to determine how much others were paying and place our ads strategically. Because ads can be set on a weekly auto-renew schedule to subtract funds from a user's account automatically, this recruitment method requires little to no maintenance.

In the first month of recruitment, we also used the Second Life Forum, a community discussion board where users can post on a variety of Second-Life–related topics to be included in the following categories: people, places and events, creation, technology, commerce, land, international, and adult. After logging into Second Life, users can access an

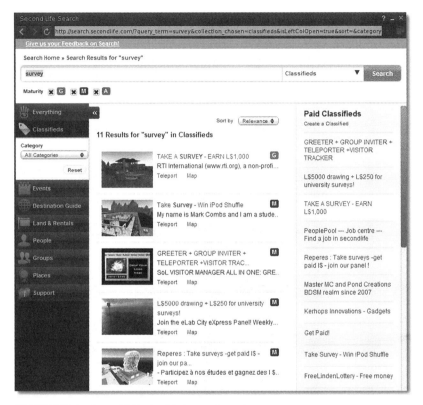

FIGURE 10.1 Second Life classifieds search results.

online knowledge base, a Second Life blog, and an answers forum. A single recruitment test was posted on December 27, 2012. Several users responded with inquiries for additional information, which the study's recruitment coordinator answered.

Additionally, word-of-mouth helped with recruitment. Researchers are aware that some participants chose to post about the study on their personal Facebook pages and Twitter accounts; however, the extent to which such posting occurred is not known.

We also recruited on the social networking sites Facebook and Twitter for this study. Messages were posted to three Second Life Facebook pages: the Second Life Games/Toys page, the Second Life Local Business page, and the Second Life Interest page. Researchers posted these messages from their accounts. Before posting, Facebook users must "like" a page. Combined, more than 270,000 Facebook users have

FIGURE 10.2   Recruitment coordinator's Twitter profile.

"liked" these pages. Messages were posted to Facebook during the first month of the recruitment period. A Twitter profile was created for the recruitment coordinator's avatar during the last month of recruitment (see Figure 10.2). Brief, 140-character recruitment Tweets were sent to user accounts or "handles." These messages asked users to re-Tweet the original message so their followers could view it. General information about the study was included as well as a link to an eligibility survey. Five Tweets were sent during our recruitment period. We are not aware that any of these messages were re-Tweeted.

A partnership established with the Second-Life–oriented blog NewWorld Notes (NWN) allowed us to recruit a general population of Second Life users using the Internet. Since Second Life was created in 2006, NWN has reported on topics of interest to Second Life users including digital technology and other augmented realities, gaming design, virtual fashion, and issues related to online identity and virtual community. Through NWN, a permanent ad was placed on the blog's website, which linked readers to an eligibility survey. Three notices were posted on the blog announcing the study. Notices also linked to posts on Facebook, Twitter, Plurk, and Google+. See Figure 10.3 for a study notice that NWN posted to Facebook.

FIGURE 10.3 Study notification on Facebook.

Targeted Recruitment

We used general recruitment methods to recruit the Second Life community at large. However, targeted methods focused specifically on our populations of interest including a referral program, information sessions held in health-related support communities, and general networking among related communities.

A referral program was initiated to boost recruitment and encourage users to refer their friends or acquaintances in both Second Life and real life. After users completed the health survey, the recruitment coordinator shared a Second Life notecard inworld that offered more information about the program. The recruitment coordinator also described the program and the notecard contents to individuals using Second Life's inworld private chat feature. This program allowed participants who had already completed the survey to refer up to three residents for additional compensation. If those referred were selected to participate in the survey and completed the health survey, study participants received an additional incentive of L$100 or approximately US$0.40.

Thirteen information sessions held in two support communities were another venue for targeted recruitment. The sessions were held in HealthInfo Island and the Chilbo Community. We felt that members of our target populations might use these communities inworld. HealthInfo Island was chosen specifically because of the health-related interests of the community. Its mission is "to provide timely, accurate, and accessible information on topics of physical, emotional, and mental health" (Second Life, 2013). HealthInfo Island boasts a health library designed to assist Second Life users with their health information needs. Additionally, the Chilbo Community includes a virtual Education Resource Center, which provides materials and resources for educators and those interested in teaching, learning, and research. Both groups were willing to serve as informal sponsors of the study and were actively involved in recruiting. The recruitment coordinator's avatar hosted information sessions in these communities using Second Life's inworld audio feature (see Figure 10.4). All sessions were also transcribed in Second Life's public chat feature to accommodate users with hearing disabilities or those without technological sound capabilities. Information sessions lasted between 15 and 30 minutes, during which time participants were allowed to ask questions pertaining to the study. At the end, attendees were given the option of completing an eligibility survey.

FIGURE 10.4   Recruitment coordinator giving presentation at HealthInfo Island.

Leaders of other support communities were also contacted to help with recruiting study participants. Communities were located by conducting specific keyword searches inworld including health, wellness, diabetes, chronic pain, HIV/AIDS, cancer, alternative health care, yoga, qi gong, reiki, acupuncture, chiropractic, and healing. In total, 84 contacts were made. Additional keywords were added to this search after researchers discovered that some members of the populations of interest might search for health information and/or support from alternative health-care resources.

Overall, community leaders were very receptive to our study and several offered to help recruit participants by messaging group members or placing notecard givers in their inworld locations (see Figure 10.5).

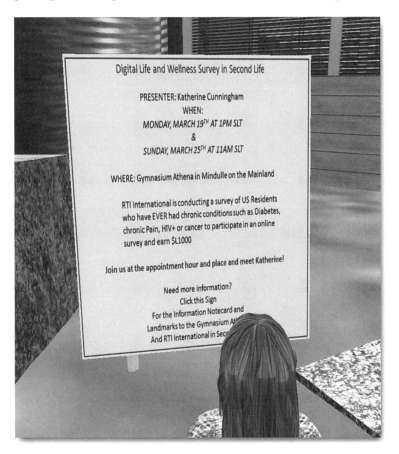

FIGURE 10.5 Notecard giver at the Chilbo community in Second Life.

Notecard givers can appear as a poster, sign, or brochure and work by relaying a textbox when engaged, usually by clicking on the sign itself. The notecards contained important information about the study and could be automatically transferred to a user's inworld Inventory. The Inventory acts as a folder, which stores the notecard to be referenced at a later date. The notecards also allowed users to share information about the study with their friends or acquaintances; however, the original information could not be modified by other users.

A notecard giver was also placed in the virtual RTI interviewing facility to inform traffic through the location about the study (see Figure 10.6).

FIGURE 10.6   Notecard giver/sign at virtual RTI interviewing facility.

Survey Administration

A brief eligibility survey was required prior to completion of the health-related survey. All recruitment materials contained a link to this web-based survey that could be accessed outside Second Life. The recruitment coordinator reviewed the results of the eligibility survey daily and contacted individuals who were eligible in Second Life. These messages were delivered through Second Life's private chat feature, viewable only to the recipient. This message notified respondents that they were eligible to participate in the health survey and provided a link. Users did not have to be logged into their account at the exact moment the message was sent; messages could be stored and replied to the next time the user was inworld. The survey could be completed at the user's convenience, and reminder texts were sent to respondents who had not completed the health survey after 7 days of the initial notification.

RESULTS

A total of 573 participants completed the screener survey, of which 236 were eligible; 181 of those eligible completed the health survey. The eligibility survey asked participants to report how they had heard about the study. Respondents could select more than one method from a list (see Table 10.2).[1] In addition to these responses, we also included some that were not used during recruitment such as "your local newspaper" and "the New York Times"—these are listed as "False Categories" in Table 10.2. These options were included to validate further accurate responding and to identify dishonest survey participants—if respondents chose one of these answer options, they were not considered eligible to participate. We also included an "Other" category, which some participants selected.

The method that yielded the highest number of recruited (210), eligible (71), and completed surveys (55) was the NWN blog. The recruitment methods that participants reported with the second and third highest frequencies were the Second Life Forum (98) and the

---

[1] Respondents were not given the option of choosing the NewWorld Notes blog (this method was implemented after data collection began) or word-of-mouth when responding to the eligibility survey.

TABLE 10.2
Recruitment Methods

	# Recruited*	# Eligible	# Completed
Second Life Classifieds	39	16	14
Second Life Forum	98	39	26
Referral Program	69	32	26
RTI Avatar	22	14	9
Facebook Post	18	4	1
Craigslist	18	7	1
Second Life Support Community	57	36	30
NewWorld Notes Blog	210	71	55
Word-of-Mouth	17	5	4
False Categories	5	1	0
Other	63	30	26
Total	616/573**	255/236**	192/181**

*Refers to the number that completed the screener survey.
**Represents the number of uniquely recruited participants. Data based on self-reports; participants could choose more than one recruitment method.

referral program (69). The Second Life Forum also recruited the second highest number of eligible participants (39) and yielded 26 completed surveys. Participants recruited from a Second Life Support Community completed the survey at the second highest rate (30).

The populations of interest were Second Life users with diabetes, chronic pain, HIV/AIDS, or cancer. Table 10.3 outlines the number recruited for each method by disease population.

These data were self-reported for all recruitment methods except the NWN blog. A separate (NWN-specific) link was generated for this recruitment method to verify respondents' self-reported responses (through the specified Other answer category) with http referrer data collected through the alternate link. Three numbers are provided in Table 10.3: the number that self-reported the NWN blog and used the NWN-specific link; those that reported they heard about the study from NWN but clicked the general link; and the total number that self-reported NWN. Another general link was used for all other recruitment methods.

The recruitment method reported with the highest frequency was the NWN blog for chronic pain, diabetes, and cancer. Respondents with HIV/AIDS (3) reported the survey referral program most. The Second

TABLE 10.3

Recruitment by Chronic Condition

	Diabetes		Chronic Pain		HIV/AIDS		Cancer	
	n	%	*n*	%	*n*	%	*n*	%
Recruitment Method								
Second Life Classifieds	8	8.8	15	7.1	2	13.3	4	13.8
Second Life Forum	15	16.5	28	13.3	1	6.7	4	13.8
Referral Program	10	11.0	28	13.3	3	20.0	3	10.3
RTI Avatar	6	6.6	12	5.7	1	6.7	0	0.0
Second Life Facebook Post	3	3.3	4	1.9	2	13.3	2	6.9
Craigslist	1	1.2	7	3.3	1	6.7	0	0.0
Second Life Support Community	12	13.2	27	12.8	1	6.7	5	17.2
NWN (Total)	27	29.7	58	27.5	2	13.3	8	27.6
Self-Report, NWN link	26	96.3	52	90.0	1	50.0	8	27.6
Self-Report, General Link	1	3.7	6	10.3	1	50.0	0	0.0
Other	9	9.9	32	15.2	2	13.3	3	10.3
Total*	91/78**		211/201**		15/8**		29/27**	

*Respondents were allowed to select more than one answer option. As a result, totals are slightly larger than the number of individually recruited participants at 78, 201, 8, and 27.

**Total number of recruited respondents by chronic condition totals 314. This is slightly higher than the total number of recruited participants reported above (236). Respondents were allowed to choose more than one chronic condition.

Life Forum had the second highest number of recruits for both chronic pain and diabetes at 28 and 15, respectively. A total of 28 respondents with chronic pain also reported the referral program and 10 with diabetes selected this method. Respondents with cancer (5) reported a support community with the second highest frequency, as did 27 respondents with chronic pain and 12 with diabetes.

Of the total sample of 573 recruited participants, 35.1% or 201 participants reported chronic pain, compared with 32.5% of the general U.S. population reported to suffer from this disease. When comparing our sample to the U.S. population for the other chronic conditions, we found that 13.6% reported having diabetes, compared with 8.4% of the U.S. population; 4.7% reported cancer, compared with 3.9%; and 1.4% reported HIV/AIDS, compared with approximately 0.4% of the U.S. population. From these figures, our recruited sample slightly overrepresented the number of participants with these four chronic conditions, which is to be expected given our targeted recruitment efforts (American Academy of Pain Medicine, 2012; Centers for Disease Control and Prevention, 2012).

DISCUSSION

Using existing Second Life communities and information leaders was an integral part of our recruitment strategy. Health-related support communities in Second Life served an important role in helping us establish the trust and legitimacy necessary to recruit Second Life users for an online health survey. Recruiting through a popular Second Life blog granted our study greater exposure and allowed us to reach a larger audience than other methods. As a result, relationship building and developing a professional inworld presence were important parts of recruitment. Other previously used methods that were also effective in recruiting chronic-condition populations included the Second Life Forum and a referral program.

Communities

We used communities a great deal in our recruiting. Many support groups for people with chronic conditions have formed organically in Second Life such as HealthInfo Island. These groups afford Second Life users the opportunity to connect with other individuals who share similar interests and health concerns. Because of the amount of information sharing that takes place, these communities have great potential to meet the needs of those with chronic conditions while positively affecting real-life health outcomes.

A primary leader or administrator typically organizes these groups, and these individuals serve as gatekeepers for the rest of the community. We developed relationships with several leaders during our study so that we were not perceived as interlopers in the communities. In doing so, we were able to host sessions at these sites to inform residents about our study and delineate the benefits of participation to group members. Members of the communities inferred our legitimacy based on our relationship with community leaders.

The community leaders served several important functions. Having the leader introduce us and our study to participants helped overcome resistance and boosted overall recruitment. Establishing relationships with community leaders also increased our ability to recruit inworld because leaders further publicized our study in the Second Life Events Calendar and sent blast notifications to group members prior to all information sessions.

Because community leaders also have relationships with one another, leaders helped us connect with other related communities. We used a snowball approach, which allowed us to reach more communities to target our populations of interest.

This method was effective in recruiting participants in our populations of interest. However, it is time consuming. The recruitment coordinator spent significant time inworld developing and building these relationships. Future studies should seek to balance recruitment goals with a project's budget and current technological expertise when considering this approach.

Using Existing Second Life Resources

Developing and building relationships with the communities were effective strategies in recruiting Second Life users with chronic conditions. However, we wanted to reach a broader population of Second Life users; therefore, we also conducted general recruitment activities. The most effective of these activities was the Second Life blog, NWN. We contracted with the writer to write a piece about our study and to put up an ad with a link to our study on their website, which was effective because of the blog's broad readership. The author of the blog also used his social network connections to advertise the study on Facebook, Twitter, Plurk, and Google+. This approach was effective because the author is well known in the Second Life community and has an established readership and set of followers on these sites.

Using NWN was very effective because of its broad readership. However, because this venue was not condition-specific, we had more respondents to the screener who did not qualify for the survey than with other targeted methods. This method also required an upfront cost to the owner of the blog but did not require significant time or relationship building by the recruitment coordinator.

Other Effective Methods

General methods that were successful in recruiting our populations of interest included the Second Life Forum. Even though we only posted one message to the discussion board during the first month of recruitment, our post was viewed 441 times and elicited 12 unique replies. The referral program was also an effective recruitment method. This success could largely be attributed to the sense of security provided by having a friend refer a study versus hearing about the study through other means—which may be more important for surveys collecting real-life health data.

General methods such as Facebook and Twitter, as used by the recruitment coordinator, were not as effective at recruiting eligible Second Life users from our target populations. One reason could be that while the conditions in which we were interested have high incidence among the U.S. population, there is no way of knowing the incidence among Second Life users because broad surveys of this sort have not been conducted in Second Life. Other reasons could be the lack of an established online presence and followers.

The Importance of the Recruitment Avatar

When dealing with populations with chronic conditions, researchers must establish trust and legitimacy with potential participants. This trust is particularly important in virtual communities such as Second Life that do not necessarily have external means of identity verification and where most users observe a strict separation between their real and virtual lives. Thus, establishing the legitimacy of the research organization and project team members is of heightened importance. Second Life users were sensitive to participant protections such as IRB approval, privacy and confidentiality, and data security. More so than studies that take place in a real-life setting, with most methods we found it necessary to advertise details regarding participants' rights, IRB procedures, and confidentiality. We were able to engender trust by building relationships with community leaders at HealthInfo Island and the Chilbo Community as well as through other well-known Second Life contacts (NWN blog). These relationships helped establish the requisite legitimacy and trust among our populations of interest.

In addition to developing relationships, we found that having a recruitment avatar that was knowledgeable about Second Life, including its culture and technical interface, was helpful. Participants looked at the avatar's public profile for information such as the avatar's creation date or "birthdate" to see how long the avatar had been inworld as well as his or her established payment history to confirm promise of a survey incentive.

Although Second Life has various types of avatars, including non-human forms, avatars completing research should adhere to the social norms of Second Life including adherence to the virtual world's dress code. Most avatars dress flashier and are more stylized in their appearance; as a result researchers should convey an understanding of the environmental and social context of the virtual environment in which they are conducting research, which includes a more casual appearance. If possible, a link to the research organization and/or eligibility study should be provided in the avatar's profile as well as links to other virtual-world research that the organization has conducted. The profile should also reflect a history of positive interactions in Second Life. Because of the incidence of spam and surveybots—or automated avatars lacking a human controller—a well-established avatar profile can improve perceptions of legitimacy.

CONCLUSION

Recruiting virtual-world study participants is very different from recruiting for traditional studies. Researchers should set time aside to gain a working knowledge of Second Life, including establishing a research avatar, before conducting research there. The recruitment coordinator's avatar should spend time cultivating and developing a series of positive interactions in Second Life. Such investments of time will help recruit people with chronic conditions as well as other research efforts.

Because participation takes place online, issues of trust and privacy are of great concern, especially when avatars are sharing information about their real lives, including private health information.

Second Life has great promise as an alternative way to reach those with chronic conditions. Virtual worlds allow people to integrate various aspects of their virtual life, which reduces the burden to participate in this sort of study. However, recruiting people in this setting has special considerations and understanding them is paramount to the success of a recruitment effort.

REFERENCES

American Academy of Pain Medicine (2012). *AAPM facts and figures on pain.* Retrieved from http://www.painmed.org/patientcenter/facts_on_pain.aspx.

BELL, D. (2009). Learning from Second Life. *British Journal of Educational Technology, 40*(3), 515–525.

BELL, M. W., CASTRONOVA, E., & WAGNER, G. G. (2009). Surveying the virtual world: A large-scale survey in Second Life using the Virtual Data Collection Interface (VDCI). *German Council for Social and Economic Data (RatSWD) Research Notes, 40*, 1–49.

Centers for Disease Control and Prevention (2012, March 14). *HIV in the United States, at a glance.* Retrieved from http://www.cdc.gov/hiv/resources/factsheets/us.htm.

CHESNEY, T., CHUAH, S. H., & HOFFMANN, R. (2009). Virtual world experimentation: An exploratory study. *Journal of Economic Behavior & Organization, 72*(1), 618–635.

DEAN, E., COOK, S., MURPHY, J., & KEATING, M. (2011). The effectiveness of survey recruitment methods in Second Life. *Social Science Computer Review.*

FOSTER, K. N. (2011). *The Second Life of social science research: An assessment of sampling and data collection methods, data quality, and identity construction in virtual environments.* Ph.D. Dissertation, University of Georgia, Athens, GA.

GORINI, A., GAGGIOLI, A., VIGNA, C., & RIVA, G. (2008). A second life for eHealth: prospects for the use of 3-D virtual worlds in clinical psychology. *Journal of Medical Internet Research, 10*(3), e21.

HEAD, B. F., DEAN, E. F., KEATING, M. D., SWICEGOOD, J. E., POWELL, R. J., & SAGE, A. J. (2012). *Recruiting virtual world users for cognitive interviews: A comparison of Facebook and Craigslist.com advertisements.* Presented at the 2012 annual Southern Association for Public Opinion Research conference, Raleigh, NC.

MURPHY, J. J., DEAN, E. F., COOK, S. L., & KEATING, M. D. (2010). *The effect of interviewer image in a virtual-world survey.* Research Triangle Park, NC: RTI Press.

NOVAK, T. P. (2010). eLab City: A platform for academic research in virtual worlds. *Journal of Virtual World Research, 3*(1), 33.

RICHARDS, A. K., & DEAN, E. F. (2012). *Gaming the system: Inaccurate responses to randomized response technique items.* Presented at the American Association for Public Opinion Research annual conference, Orlando, FL.

Second Life (2013). Healthinfo Island. Retrieved January 10, 2013, from http://world.secondlife.com/group/a43aa3cc-e9fa-1c70-da17-d2977133f178?lang=en-US.

SIDDIQI, S., MAMA, S. K., & LEE, R. E. (2011). Developing an obesity prevention intervention in networked virtual environments: The international health challenge in Second Life. *Journal of Virtual World Research, 3*(3), 26.

SWICEGOOD, J. E., HAQUE, S. N., DEAN, E. F., RICHARDS, A. K., & HEAD, B. F. (2012). *Recruiting special and hard-to-reach populations in Second Life study: Recruiting through the eyes of an avatar.* Poster presented at the 140th annual meeting of the American Public Health Association (APHA), San Francisco, CA.

CHAPTER 11

Gamification of Market Research

Jon Puleston, *Global Market Insite, Inc.*

This chapter explores the use of gamification in social science research as a means of designing more effective research studies. Gamification has increased rapidly in market research and other areas in the last few years, and survey researchers outside the market research world are beginning to adopt it—but what & does it mean? Does gamification have any substance, and how can it be used in research? Gamification is the process of applying game-design thinking to nongame applications and activities to make them more enjoyable and, as a result, increase participation in the activity. Put more simply, it is the realization that people may more willing to participate in activities if they are fun!

Gamification is affecting many aspects of the marketing, design, and entertainment industries. Commercial enterprises use games to motivate individuals to do all kinds of activities, from exercising to shopping. Even governments are using games as a means of social engineering and revolutionizing education.

According to a 2011 Gartner Research Report, by 2015 more than 50% of organizations that manage innovation processes plan to gamify those processes (Gartner, 2011). The trend has gained momentum and support from industry heavyweights such as Bing Gordon, Al Gore, and

Social Media, Sociality, and Survey Research, First Edition.
Edited by Craig A. Hill, Elizabeth Dean, and Joe Murphy.
© 2014 John Wiley & Sons, Inc. Published 2014 by John Wiley & Sons, Inc.

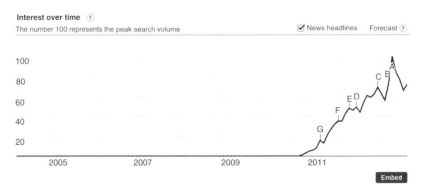

FIGURE 11.1 Interest over time in gamification.

J. P. Rangaswami (Chief Scientist of Salesforce.com). It is one of the most discussed trends in Silicon Valley, and Google Trends shows the explosive growth is continuing to accelerate (see Figure 11.1)

SIGNIFICANCE OF GAMIFICATION IN MARKET RESEARCH

The theory of gamification is particularly relevant to market research because surveys straddle the dividing line between work and fun. Completing a survey is, in most cases, a voluntary task done for altruistic reasons. Sometimes the offer of a small incentive is an additional motivator. Civic duty (in the case of government-sponsored surveys) is another example of a motivator. But these motivators are often not sufficient to result in the collection of quality data. Once respondents click on a link to complete a web survey, the amount of time they spend completing it is completely up to them. Respondents often rush through a survey to complete it quickly without thinking a great deal about their answers. In many instances, respondents are not motivated to spend sufficient time answering questions. Completing surveys is often boring because they are frequently dominated by uninspiring repetitive questions, as the example in Figure 11.2 illustrates.

When respondents encounter such questions, they usually answer them quickly so they can finish the survey as rapidly as possible. This type of behavior is commonly termed "satisficing" and defines the efforts by respondents to shorten the experience. Global Market Insite, Inc.

What topics in marketing and advertising research will be "hot" in the future? Please indicate which category best describes each item below, looking ahead 5-10 years from now. If you are not familiar with the term at all, or do not feel you know enough to evaluate it please just indicate that below.

Marketing research on/for ...	Fading fast: its days are numbered	Trendy: here today, but maybe gone tomorrow	Timeless: this is tried and true	Cutting edge: we'll be seeing more of this	Hard to say: I've heard of this but don't know enough	No idea: I don't even know what this means
Media planning/mix optimization	O	O	O	O	O	O
Mobile marketing	O	O	O	O	O	O
Digital marketing	O	O	O	O	O	O
Multicultural marketing	O	O	O	O	O	O
Neuroscience	O	O	O	O	O	O
CRM	O	O	O	O	O	O
Cultural trends	O	O	O	O	O	O
Customer loyalty	O	O	O	O	O	O
Integrating "design thinking" into marketing	O	O	O	O	O	O
Shopper Insights	O	O	O	O	O	O
Product/service innovation	O	O	O	O	O	O
Touchpoint effectivenes	O	O	O	O	O	O
Impact of social networks	O	O	O	O	O	O
Brand strategy/management	O	O	O	O	O	O
Marketing/advertising effectiveness	O	O	O	O	O	O

FIGURE 11.2   Example of typical survey questionnaire.

(GMI) has measured the average time respondents spend answering different types of questions in surveys by consolidating the average answer times from thousands of respondents across hundreds of questions. We found that on average, respondents will spend around 4 seconds thinking about their answer to a typical single question, but this answer time drops to under 2 seconds when respondents are presented with a repetitive grid and can drop to as low as 1.5 seconds if this question is presented at the end of a survey.

If survey designers can improve the survey experience for respondents—that is, make survey completion more fun and entertaining—respondents may pay more attention and put more effort into completing the survey. They could provide responses with more thoughtful feedback and higher quality—these benefits are the appeal of gamification. In some experiments conducted in association with Bernie Malinoff from Element 54, researchers investigated the relationship between how fun/entertaining questions were to answer, how much time respondents spent answering these questions, and the quality of data these questions produced. They observed a strong underlying relationship between fun and the quality of feedback, as Figure 11.3 illustrates. We designed a series of questions in increasingly more fun and engaging styles and found them to improve consideration time and reduce the level of straightline answers.

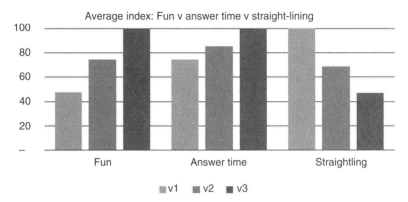

FIGURE 11.3 Average index: fun vs. answer time vs. straightlining.

Apply Gamification to Market Research

To apply gamification techniques to market research, a good starting point is to understand some of the underlying theory. What are games and what makes them fun?

What Is a Game?

The official Wikipedia description of a game is "structured playing, usually undertaken for enjoyment," but a game can be defined as any form of thinking or physical activity people do *voluntarily* for fun (Wikipedia, 2012). One's mindset is really all that differentiates a game from work. People must work, but they may *choose* to play a game if they wish. In the actual task involved, sometimes there is little that differentiates work from a game, and they can often be one and the same. A good example of the blurry boundary between work and games is the classic scene in Mark Twain's novel *The Adventures of Tom Sawyer*, where Tom has to paint a fence; he refashions this task into a fun activity and persuades his friends to do it voluntarily. Another example might be a mathematics lesson in which one has to solve a problem. If that same problem is presented in a computer game, solving it is usually fun, and some people will choose to do it. As children play video games, they spend hours solving problems almost identical to the conceptual mathematical problems they solve at school—but one is fun and one is not. Identifying what makes a task fun is at the heart of gamification.

Certain ingredients, however, make games more fun and can be used as the building blocks to construct games.

Constructing Games

1. **Rules**. Rules are the key ingredients used to redefine tasks into games, and they are what make games fun. For example, if one is asked to carry a 20-pound bag for 2 miles, the task seems like work. If, however, some enjoyable rules are applied to this task (e.g., in the 2 miles try to hit a small white ball into small holes with a minimum number of strokes), the task changes to an entertaining activity that people actually pay to play—known as golf. Another example is to make a game out of a mundane task. Walking to the store to shop becomes more fun for a child if the parent asks the child to walk without stepping on the sidewalk lines; the walk then becomes a game.

   Another important consideration is that, often, the more absurd the rules, the more fun the game is to play. Sports like football and English cricket are good examples of games with a long list of outwardly incongruous, restrictive rules that help fashion the games into interesting experiences for the player and the viewer.

   Such scenarios play out in real life: Imagine an online forum where people could chat with one another—in the early days of the Internet, thousands of these forums emerged. Then, an ingenious individual created a forum with a seemingly silly rule: only 140 characters allowed in each post. This rule instantly turned the forum into a clever game where users were challenged to condense their thoughts or opinions into just a few words— otherwise known as Twitter.

2. **Balance of skill, effort, and reward.** Many people play games at no cost or even pay to play them; the rewards are usually mental satisfaction, not monetary gains. Understanding the value of these small mental rewards helps in creating games that people are motivated to play.

   Games without some sort of skill and effort tend not to be as challenging as those that involve an effort on the player's part. If one wins a game or a competition easily, there is less of a sense of achievement. Game designers have discovered that when designing levels of games, for example, there is a sweet spot between making the effort too difficult and too easy. The combination of reward and effort has to be balanced carefully. The mental rewards for completing a difficult task successfully are greater than for accomplishing an easy task—but if the balance is not right or if it is too hard, people are discouraged from trying.

For example, if one is asked to guess how many legs a dog has, the task is too easy and, therefore, not fun. If one is asked to guess how many legs a millipede has, the task may be too difficult, with a lower chance of winning and so not as fun. But if one is asked whether a millipede has more or fewer than 100 legs, the question presents a balanced chance of winning or losing and becomes fun.

3. **The element of randomness.** Popular games often blend a certain skill with some level of randomness or luck so everyone has some chance of winning. Card games and backgammon fall into this category; much skill is needed to win consistently at backgammon, but even a first-time player with the luck of the dice can win. Again, the level of randomness has to be balanced carefully. If the mix leans too heavily toward randomness, a game can rapidly lose its appeal; the children's game Chutes and Ladders (known as Snakes and Ladders in the United Kingdom) is a good example. Most people older than 10 years of age tire of playing this game because of the random element involved. The other extreme is the game of chess, which requires a great deal of skill and knowledge; many people are discouraged from playing it after they lose the first few times.

4. **The spark of imagination.** The imagination and the thinking process are generally the heart of most games. The prefrontal cortex of the human brain is designed to imagine and think, and people enjoy using it! All games—from Eye Spy to Scrabble or video games—require concentration and often involve extremely high levels of mental processing. Human beings like to conjecture and plan scenarios; games provide an opportunity to do so (for example, the success of Monopoly is based on scenario planning).

5. **Motivation to play games.** Human beings are genetically primed to play games. Humans use games as a means to practice the skills

needed to survive and reproduce, and this instinct has been honed over millions of years of evolution. Even animals have embedded instincts to play, as anyone who has seen a kitten with a ball of yarn or a dog with a stick can attest. Scientists studying gorillas observed that they play competitive games just like humans and even solve puzzles for fun (BBC, 2010).

Humans have a near-Pavlovian response to challenges; we often respond and act without even thinking. Men and women originally needed this instinct to survive. Imagine one of our ancestors in the wild who suddenly saw a movement; he did not know if it was food or a predator. He responded automatically without the need to think. Today, this instinct to respond to a challenge or perceived danger still often occurs immediately. If a parent tells a child he will race her to the car, she starts running. If that same child is about to run into a busy street, her parent will instinctively race to place himself between the traffic and his child. Humans are motivated to do nearly anything if there is a clear danger—or a fun purpose. A task can be converted into a game-like experience simply by finding a motive for doing it—often, the opportunity to defeat someone else in a challenge or even to improve one's personal record is sufficient.

Gamification in Survey Design

Exploring the Role of Gamification in Research Context

Over the past 6 years, I have been exploring ways to improve the feedback from online surveys by making them more engaging for respondents. I have explored many techniques and conducted over 100 research-on-research experiments. Nearly all of this work has been done in partnership with research companies, academic institutions, and end-user clients, specifically Deborah Sleep from Engage Research with whom much of the earliest pioneering work was conducted; Bernie Malinoff from Element 54, a prominent researcher exploring research design issues; Duncan Rintoul from the Institute for Innovation in Business and Social Research, University of Wollongong in Australia, an expert on research theory; Mark Utley, former head of research at Sony Music UK (now working for Bloomberg); Daniel Hall, current head of research at Sony Music UK; and Mitch Eggers, GMI's Chief Scientist.

Over the course of these experiments we learned that the design and ergonomic flow of surveys were crucial and that the dropout rate could be reduced by making surveys more engaging consumer experiences. We also learned the value of imagery in online surveys as a means to stimulate the imagination, trigger the memory, and encourage more enthusiastic participation in a survey, especially when an element of humor was used in the imagery. In partnership with Deborah Sleep, we examined how to improve the language used in surveys, and we saw how, when the tone of wording was friendlier, a better relationship with respondents developed, which encouraged more feedback.

We applied thinking borrowed from social psychology and qualitative research and discovered how dramatic an impact such techniques as projection and imaginary scenario planning could have on the respondents' willingness to complete survey tasks, and on the effort they put into the tasks. With various partners, we also tested more creative questioning techniques and found that when we made questions more game-like, click count improved, straightlining effects were reduced, and respondents' satisfaction levels increased.

We also noticed the impact that humor had in surveys: Segueing questions with a humorous animation effect seemed to reset respondents' concentration levels and improve the feedback to follow-on questions. We observed that introducing any fun or game-based mechanic tended to provoke positive reactions from survey respondents. For example, in one experiment conducted with Engage Research, the use of the phrase "we challenge you" stimulated a threefold increase in the number of ads respondents recalled. In another experiment, applying a 2-minute time limit (to respond) resulted in 10 times as much feedback as the previous version of the question. In a third experiment, incorporating the words "can you guess" extended the time respondents spent considering a question from 10 seconds to 2 minutes. Even small changes to question design, such as having emoticon stamps appear when the answer was clicked, seemed to encourage greater consideration in response choice.

When we asked respondents why they had written more and spent more time thinking when answering these types of questions, their answer was simple: It was more fun! We slowly discovered the impact that gamification could have on improving response levels to surveys.

Adopting the Concept of Gamification

As a result of this discovery, we decided to explore the idea of game-play in greater depth. We examined the theory behind game-play and how it was being used in other fields, with the aim of discovering how we could integrate this thinking into surveys.

The concept of gamification, we quickly discovered, was already sweeping across the marketing communication industry and was being discussed in marketing departments, advertising agencies, and even governments around the globe. We found countless examples of how it was being applied to encourage compliance and active participation in all manner of different activities.

Market research had great potential to incorporate gamification successfully because many surveys could already be seen as games, albeit boring, badly designed ones. GMI then embarked on a series of groundbreaking experiments exploring the role of gamification; we began by rewording questions to make them more fun and then examining what impact the rewording had on the data and on the construction of whole survey game experiences. The rest of this chapter explains what we learned from these experiments and summarizes how gamification can be used to improve survey design. The original experiments are catalogued in a series of papers (cited in this chapter's references); the two most notable are *The Game Experiments* (Puleston & Sleep, 2011) and a follow-up paper, *Can Gaming Techniques Cross Continents?* (Rintoul & Puleston, 2012a).

Rethink Question Writing

Respondents often view surveys as boring because of the questions themselves and the way they are usually asked—the questions can be dry and emotionally uninvolving. To make surveys more game-like (so as to increase rate of response), survey designers must rethink the style and approach to writing questions.

The language used in many surveys dates back to the era of face-to-face interviewing techniques where the need to engage respondents was not as significant an issue. So as not to bias responses, question writers regarded clear, understandable language and presenting ideas as neutrally as possible as the most important factors. Without any visual cues, survey language also had to be very descriptive.

This history often led to the adoption of verbose, overblown phrasing of questions based on the desire to explain every nuance of the task at hand, despite the knowledge that simpler is better; somehow, this complexity was established as the accepted style for writing surveys. (I read through hundreds of surveys every year, and one of the first problems we have to resolve is overly verbose descriptions of instructions for respondents—instructions that I know from experience many respondents will not bother to read.) A classic example is the wording: "on a scale of 1 to 10, how much do you agree or disagree with these statements where 1 means you completely disagree and 10 means you completely agree and 5 means you neither agree nor disagree." If the respondents see a scale with 0 at one end and 10 at the other with explanatory labels, a descriptive passage in the instructions is unnecessary and, indeed, can make reading survey questions seem more like poring through a legal document. All this work can demotivate respondents from actually reading instructions properly. With online research it is crucial that a more succinct, fun, and engaging approach be employed.

In interviewer-administered surveys, the survey administrator is usually with the respondent, and there is less of a concern about whether the respondent is actually listening to the questions or answering them properly because the interviewer is physically there. He or she can see, at least to some extent, how closely respondents are paying attention.

Rarely during interviewer-administered surveys would one expect respondents to walk away halfway through an interview without responding to any questions or to give the same answer repeatedly to a set of questions, and an interviewer is unlikely ever to read just the first line of every question. But with a computer (or any self-administered mode), respondents can all too easily give just a fleeting thought to the questions, not read them properly, make up answers, choose any option, and click the "Next" button quickly—or just close the survey if they get bored.

GMI has conducted research on how long respondents spend reading and answering questions. This research found that, in a 15-minute survey, up to 50% of respondents do not read question instructions properly, and only around 15% of respondents gave due consideration to every single question. In our experiments, 85% of respondents "speed-answered" at least one or more of the control questions; speed-answering means responding in half the average time—the classic measure by which "speeders" in surveys are defined. Nearly 40% of respondents have been found to stop responding to an online survey halfway through if they consider the questions boring.

Change Question Style

Gamifying a survey must begin by considering a more engaging approach to question wording. Question writers can use the following recommended techniques to encourage respondents to read and answer questions.

1. **Personalization.** Make the questions relate to the respondents, as if "it is all about them." This powerful technique encourages people to respond; most are fundamentally more interested in answering questions about themselves than other topics. When respondents choose words from a list they associate with a brand versus choosing words they associate with themselves, they will spend up to two to three times longer thinking about their answer when the question concerns them: This effect is an indication of higher engagement.

   Thus, the question "Which of these paint colors do you like best?" could be personalized by asking "If you had to paint your room in one of these colors, which one would you pick?" Or, as another example, instead of asking "How much do you agree or disagree with these statements: I like cheese; I believe in capital punishment, etc.?" you might ask "How much like you are the people who made these statements? I like cheese, I believe in capital punishment etc." Research has shown that making questions more relevant to respondents can increase the time respondents think about the answers by up to 30%.

2. **Emotionalization.** Respondents' latent feelings can encourage them to think more about the question. For example, the question, "What would you wear?" could be revised to read, "What would you wear on a first date?" This form of personalization targets emotion, and an emotional trigger in a question can significantly improve consideration time and the quality of feedback from respondents. As an example, in a GMI-conducted experiment, one group of respondents was asked to list the foods they disliked eating; for the second group of respondents, researchers emotionalized the question by showing a picture of a person

expressing disgust. In the second group, the number of foods listed doubled from a mean average of 3 to 6 per respondent.

3. **Projection**. This technique is used frequently in qualitative research; respondents are asked to imagine something in the mind's eye of someone else (this powerful engagement technique can be used in online surveys too). As an example, a question might be "What do you think about this new product?" Change this to "Imagine you are the boss of a company. Your job is now to evaluate this new product." GMI has conducted various experiments using this technique and has found that this technique can easily encourage up to twice as much attention and feedback from respondents when used the right way. Some of these experiments were catalogued in a presentation for to the Advertising Research Foundation (Sleep & Puleston, 2009).

Figure 11.4 is an example of an experiment in which respondents were asked to give critical feedback about an advertisement. They were asked to imagine they worked for an advertising agency and a competitive company had released a new commercial; they were to be shown

FIGURE 11.4 Experiment: critical feedback to advertisement.

a sneak preview and report what they thought of the advertisement. The result? The volume of feedback increased from a mean average of 14 words per respondent in the control cell to 53 words in this test group.

Apply Rules to Question Design

Rules can be used to transform questions into more game-like experiences for respondents; rules can be used to turn questions, in effect, into mental puzzles that make them more interesting to answer.

Adding rules to a question is often done quite naturally, to the extent that people do not realize rules are there. In fact, a question is a rule in itself. Effective rules should be imaginative, abstract, and in certain circumstances, irrational, to respondents. Such wording can challenge their perceptions, which is a good trigger for engagement. For example, one could ask, "What brand do you like?" or "What brand would you want to wear on your T-shirt, or have as a sponsor of your football team?" These methods of measuring liking are more interesting for respondents to answer and can enliven a dull survey.

The following example, taken from one of the first game experiments GMI conducted (Puleston & Sleep, 2011), shows how rules can be added to questions to make them more game-like:

- Question: "Please describe your favorite meal."
- Question with rule: "Imagine if you were on death row and had to plan your last meal. What would you order?" (i.e., the rule is that it is the last meal).
- Question response : A mean of 3 words per respondent.
- Question with rule response: A mean of 15 words per respondent.

Figure 11.5 shows the difference in the responses to the first two questions; note how much more information was divulged when the question was gamified.

Rules can be applied to questions in many different ways including the following techniques that work well.

1. **Make the question specific.** Adding a specific scenario to a question is the easiest and most versatile rule technique to use. Many question authors use this technique quite naturally; take the

Favorite meal	Death row Last meal
Steak au pouivre	Scotch broth soup as a starter served with garlic bread. Medium grilled gammon steak with a lightly fried egg on top with chips and side salad. A glass of red wine. A sticky toffee pudding, followed by cheese and biscuits.
Pesto Pasta	
fish and chips	Bacon chips and tinned tomatoes and an egg butternut squash soup(homemade) fillet steak well done (not burned) new potatoes fresh garden peas fresh fruit salad and cream all served with a good white wine
garlic chicken	
	Calamari for starters Curried goat with rice & jerk chicken with plantain and fried dumplings for main meal and hot sticky toffee treacle pudding with hot custard for dessert with a triple amaretto and lemonade to drink
	Classic roast dinner with beef cooked medium rare, pink in the middle and a little bit of blood, the roast potatoes crispy on the outside soft on the inside, Yorkshire puddings, peas and gravy cooked in shallow oil and chips

FIGURE 11.5 Application of gaming rules to survey questions.

example of a question about clothing ("What would you wear?"). The question is more interesting to answer if it is written in a specific context:

- What would you wear last year that you would not wear this year?
- What would you wear if it were a hot, sunny day?
- What would you wear if you were going to a job interview?

The key in this technique is to tailor the rule to the context you wish to understand.

2. **Use restrictive rules.** In this technique, the author limits what respondents can do or how they can respond, e.g., restricting the number of words respondents can write or select, which can have the opposite effect from what is expected.

If people are told "Describe yourself," on average, they write 2.4 descriptors about themselves; however, if the item reads, "Describe yourself using only 7 words," an average of 4.5 descriptors result—twice as much feedback as without the restriction. In addition, respondents spend more than twice as much time answering the question when it is phrased this way.

LinkedIn discussion updates illustrate other examples of successful use of this technique. The number one conversation in the

Innovation and Entrepreneurship Society LinkedIn group, with over 176 comments, is "What do you do in EXACTLY 7 words?" The number one conversation of the ESOMAR LinkedIn community is "Define talent in one word" with over 201 comments (and counting).

The same is true with word list selection. Instead of saying "Pick the words you associate with this brand," say "Describe this brand in no more than 5 words using the list of words below." Psychologically, this wording is more like a game and it encourages people to thinking for a longer time; the result is that they will actually choose more words than they would without the restriction.

3. **Whittle down rules.** Forcing respondents to make decisions is another effective way of asking questions (see the following examples):

- In your clothing collection, which is the best article of clothing and which is the worst?
- If you were planning a vacation and could only take three outfits, one for daytime, one for evening wear, and one for a special occasion, what articles of clothing would you take?

Motivation: Turn Questions into Quest!

Games must have a clear, entertaining purpose to succeed. This same thinking can be applied directly to question wording and survey construction. Often, little thought is given to what motivates a respondent to answer a question or a survey, but it is a central question to consider in gamifying a survey.

Imagine the evolutionary roots of game-play—the propensity to play games evolved to help human beings hone and develop their survival of the fittest skills. When a kitten plays with a ball of yarn, it is practicing to kill prey (Maestripieri, 2012). Prehistoric humans went on extended gathering missions for several days during which time our brains rapidly developed to gather and store information to maintain extended periods of concentration and eliminate all other distractions (The Independent, 2012).

Many children's games (hide and seek and various sports) provide an opportunity to practice these basic hunter-gatherer skills by encouraging the players to focus and concentrate. The best games guide people directly into this hunter-gatherer mindset. Consider some of the most successful video games—*Call of Duty* or *World of Warcraft*—that encourage players to spend hours at a time doing quite mundane tasks to complete a quest. This same thinking can be applied directly to question design. By rewording questions to seem more like quests and missions, the amount of time respondents are prepared to spend answering them can be significantly increased. We have conducted numerous experiments exploring this idea, and this thinking can be applied at both a question and a survey level.

A question such as, "How much do you like these musical artists?" could be reworded to read, "Imagine you owned your own radio station and could play any music you liked. Which of these artists would you place on your playlist?" The number of artists respondents were prepared to rate increased from 84 to 148. The wording made the question more relevant to respondents—it had a point and it became a quest. As a result, respondents focused more on the task and spent twice as much time doing it.

This thinking explains why, for example, projective questioning techniques are so successful (e.g., the projective experiment cited earlier where respondents were asked to imagine they worked for an advertising agency and were about to see a sneak preview of a TV campaign from a rival agency for a rival brand). This experiment set up a quest for respondents, and the volume of feedback increased more than fourfold. This versatile technique can be applied to all manner of different survey question circumstances.

The concept can be scaled up to apply not only to individual questions but also to entire surveys. For example, instead of asking people to rate their "mobile phone services provider," the question could begin "Imagine your mobile service provider was on trial and you are the judge."

Another example from the original game experiments conducted by Puleston and Sleep (2008) started with a typical media TouchPoint survey where people were to answer a repetitive series of questions about awareness and use of different media. To "questify" this survey, they asked respondents to "Imagine you work for an advertising agency and have to plan an advertising campaign with £1 million budget to reach a really important target audience . . . people like you!" Respondents were given an almost identical version of the original TouchPoint survey and the quality of responses improved significantly.

- Twenty percent of respondents spent more time answering like-for-like questions, compared with traditional control media TouchPoint surveys.
- The average selection of the "no exposure to media" option dropped from 23% to 14% (an indirect measure of the depth of thought respondents devoted to answering the question).
- About 82% of respondents enjoyed the second version, while only 61% of respondents enjoyed the first version, even though the second version took 4 minutes longer to complete.
- The average completion rate increased from 78% to 95%.

Scenario Planning

A part of the prefrontal cortex of the human brain has specifically been developed to plan, and many of the games we play centered on scenario planning, e.g., "what if?" and other imaginary situations (Koechlin et al., 2000). Changing a question to evoke a process more like scenario planning is another way of making questions more game-like for respondents. Instead of asking "What words would you use to describe this brand?" the question could ask "Imagine this brand was a human being; what words would you use to describe this person?" We have found that this twist in wording encourages respondents to name up to twice as many characteristics.

Add the Competitive Element

Most games involve a competitive element, and adding any form of competition to a survey seems to provoke a strong positive reaction to the task.

Impose Challenges and Time Limits

Instead of asking what brands of insurance company come to mind, ask "How many insurance companies can you name?" Respondents will

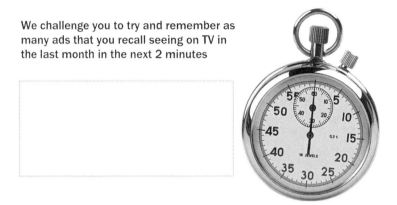

We challenge you to try and remember as many ads that you recall seeing on TV in the last month in the next 2 minutes

FIGURE 11.6 Time limit encourages increased responses.

often list twice as many brands. Adding a restrictive rule in the form of a time constraint to this process seems to encourage even more responses (see Figure 11.6).

If, for example, respondents are told they have only have 2 minutes to answer the question, they think they do not have much time and they rush to provide answers. They treat the time restriction as a challenge. In one experiment in which respondents were asked to name foods they like to eat, the number of responses increased from an average of 6 foods to 35 when the time constraint was imposed.

The reality is that a 2-minute time limit is not a constraint at all. Respondents might typically spend only 30 seconds answering this type of question; however, by stating the 2-minute time limit, respondents are actually encouraged them to spend four times as much time answering the question—which they will invariably use.

Ask People to Guess What Other People Think

Asking people to guess what other people think or have said is another powerful technique that encourages responses as a competitive challenge. Thus, instead of asking people to pick the words they associate with a certain brand, the question asks people to guess the top five words people associate with the brand. Although this variation can change the answers, it can also easily encourage respondents to spend twice as much time thinking about their answers.

This type of question is quite effective at the beginning of a survey to get respondents thinking about a topic under research and to open a discussion about it. For example, GMI conducted research for an insurance company (obviously quite a dry topic), so the survey authors began the survey with the question, "We asked 100 people what emotions they felt when they heard the phrase 'your car insurance is up for renewal.' Can you guess the most common things they felt?" This question was followed by a phrase such as, "That is interesting you selected xxxx. Many other respondents picked this response too. Can you explain why?" This statement let people know how they compared with other "players" in the "game," and it opened a dialogue with respondents, giving them some feedback, a sense that their responses were being listened to, and they meant something. This type of dialogue is like a chat room, and it got people interested in a topic that was not very exciting.

Add Reward Mechanics

People usually play most games to achieve or earn a reward, but these rewards are mostly mental: the positive feeling you get for giving a correct answer or from winning more tangible rewards such as points and prizes for succeeding. Adding reward mechanics into surveys is one of the most powerful means of gamifying a survey experience and can be accomplished in several ways. GMI has explored the option of giving respondents points for making correct predictions and trading style games where respondents are asked to trade, invest, or gamble virtual points or virtual money on their own opinions.

Both techniques can induce twice as much time spent on a task and significantly improve survey enjoyment levels. As an example, again taken from GMI's original game experiments, interviewers asked respondents to predict the future of different brands (Figure 11.7). To make this process more game-like, interviewers gave each respondent a trading budget and asked them to gamble on their opinions. If the "market" agreed or disagreed with their opinions, they won or lost the money they had gambled.

Using this mindset, respondents easily spent more than twice as much time thinking about what answers to give (see Figure 11.8).

This technique effectively eradicated straightlining (people clicking the same answer for every brand) in the control experiments. Without the technique, 18% of the 200 test respondents straightlined; under

What is your prediction for Apple and how much do you want to bet?

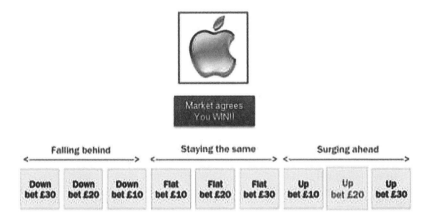

FIGURE 11.7 Predicting the future of a brand name.

FIGURE 11.8 Respondent average consideration time (in seconds).

the experimental condition, this figure dropped to 1% of the 200 control respondents. When respondents asked about the experience of answering survey questions in this way, their feedback was almost universally positive. Over 90% said they found it fun compared with fewer than 30% in the control group.

Give Feedback

Simpler rewards can consist of just giving feedback. For example, if a project called for some segmentation work (a common market research technique in which people are asked a battery of questions to determine what groups they can be classified into), interviewers could tell people at the end of the set of segmentation process roughly what segments they were in. People find this information interesting, and it gives more purpose to doing the task. Just the simple process of telling people how other people answered a question is a great way of giving feedback. If respondents are encouraged to be more curious about what other people say, they may think more about their answers. This scenario is the essence of a game—players take an action or give a response, and then they see the results.

In some circumstances, interviewers would not want to use this technique because it can encourage people to adapt their answers. For example, if people were asked to report how well they do certain things, telling them they will find out at the end how they have done may encourage them to exaggerate. Bear this possibility in mind when considering the use of this technique. Under the right circumstances, it has proven to be a successful means of engagement and a way of improving the quality of feedback. The more feedback one can integrate into a survey, the higher the level of consumer satisfaction, which is linked directly to the time and effort respondents are prepared to invest in completing the survey.

Make Tasks More Involving

Games often involve a considerable amount of mental processing. People like to think. Nearly all the game-play mechanics discussed so far actually make the process of answering questions slightly harder by inserting hurdles that must be overcome. The most popular games are successful because nearly all of them call for more thought and effort than is normally expected of people when they complete surveys—games such as Scrabble, Chess, or most card games are good examples.

Survey question designers are often so fixated with asking respondents simple questions such as "What is your favorite color?" and "How much do agree with this statement?" that they do not encourage a great deal of thinking. GMI's experience has shown that the more complex the task, the more fun, enjoyable, and rich an experience the survey is for respondents. For example, in the advertising experiment outlined earlier, one group of respondents was given a complex media planning task, where they had to plan an advertising campaign and weigh the

You can buy up to 10 advertising exposures in any combination of media you choose but you cannot spend over £1m (prices here are quoted in £000's).
Build a media plan by dragging any media into the boxes below.

				Budget: £1,000.00		Total: £0.00		

Daily Newspaper ad £150.00	Free newpaper ad £100.00	Local'/regional newspaper ad £70.00	Monthly Magazine ad £150.00	Weekly magazine ad £90.00	Live music event ad £40.00	Commercial Radio ad £90.00

Cinema ad £400.00	TV ad £300.00	Interet ad £40.00	Social media ad £60.00	Online music site ad £30.00	Online games ad £40.00	Mobile aps ad £40.00

Mobile internet ad £20.00	Bus advert £80.00	Large roadside poster £40.00	Small steet poster £60.00	Underground poster £120.00	Direct mail £40.00

1	2	3	4	5	6	7	8	9	10

FIGURE 11.9   Example of complex media planning task exercise.

relative value of various media. This complexity added an extra 4 minutes of time respondents took to complete the survey (see Figure 11.9).

Despite being more work for respondents, they actually rated their enjoyment of the more complex survey 10% higher than the companion survey.

Another example is an ethnographic research study GMI did for Sony Music. We added a rule asking respondents to imagine they were being interviewed by a magazine and that what they wrote would be published. The addition of this rule resulted in respondents spending 30% more time answering the survey and the overall word count of their answers increased from a mean average of 230 to 350 words. The task was far more interesting when reframed this way.

When constructing questions, survey designers are often overly concerned that not everyone will be able to answer more complex questions, so questions tend to be directed to the lowest common denominator. The thinking is that everyone should be able to answer every question. This perspective should be challenged. As long as more complex tasks are completely voluntary, many people are apt to respond to more complex and thoughtful tasks in a survey. Think of the people who might complete surveys as 200 consultants (rather than as 200 unintelligent respondents): All have unique and intelligent points of view that can be used to help solve a problem.

Ensure the Challenge Can Be Accomplished

For games to be successful, players need to view them as a challenge that can be accomplished with the right balance of effort, luck, and skill. This consideration is also important in the context of a survey. Suppose, for example, one made a brand awareness survey into a game called "Guess the Brand." Players/respondents would be shown facets of the brand's identity, perhaps part of the logo, a strapline, or an extract of an advertisement, and they would be asked to guess the name of the brand. If the logo happened to be the McDonalds "M" logo, most people know that logo and would not derive much challenge or pleasure from guessing it. Conversely, if people were shown the logo of a really obscure brand and asked to guess it, they may not even be aware of the brand itself, which is frustrating. Computer gaming communities aim for a "win" rate of 25% normally, but they expect people to play these games more than once; if they were too easy, gamers would not want to play again. In the context of surveys (usually a one-time experience), the win rate must be higher. Depending on the number of game mechanics to be implemented, a reasonable rate of "wins" (or expected successful attempts at whatever the game may be) is between 50% and 80%. Although I do not have evidence to confirm this rate, my experience suggests this level will result in the right amount of engagement with the survey task. Before implementing, though, I suggest testing your game on friends and colleagues to get some objective feedback on the difficultly of the task and just how much fun and engagement it added to the survey.

HOW TO DESIGN QUESTIONS TO BE MORE GAME-LIKE

How questions are designed has a powerful impact on making surveys more fun and game-like. Survey technology somewhat limits designers' creativity, but the following fundamentals are essential.

1. **Use more imagery.** Most computer games are enormously visual experiences; for a survey to become more gamified, imagery

FIGURE 11.10   Images versus no images in a survey question.

plays a vitally important role. Research that GMI has conducted over the last few years has shown that respondents greatly prefer answering more visual-based questions.

Multiple-choice questions using images instead of text generate more clicks and more thought and effort put into the selection. In the questions shown in Figure 11.10, an average of 23% more clicks were recorded when images were used.

During one experiment, two versions of a survey were created: one with imagery and one without it. GMI compared the dropout rate and the propensity to take part in a follow-up survey and had a detailed, up-close look at the data generated. Thirty percent more people completed the survey with imagery, and in like-for-like questions, responses increased by up to 50%. Also, 25% more people volunteered to take part in a second wave. All these factors meant that the volume of feedback doubled for certain questions.

Imagery encourages respondents to stop and think. For example, a picture of a water faucet was added to a set of questions about drinking water (see Figure 11.11). After the image was added, about 35% more incidents of drinking water were recorded during the daytime hours. Note: We are not making the assumption that, in all cases, more responses are better; however, in most cases, researchers are hoping for higher volumes of responses.

2. **Challenge the layout and design rules.** Most survey technology came from pen-and-paper thinking with rigid grids and frame structures and a limited visual repertoire or graphical design interface. One of the best ways to make survey questions more

When did you drink some tap water yesterday?

Please click with your mouse on all the times you recall drinking tap water

	7am	8am	9am	10am	11am	12am	1pm	2pm	3pm	4pm	5pm	6pm	7pm	8pm	9pm	10pm	11pm	12pm
Drink tap water																		

I did not drinik any tap water

FIGURE 11.11 Image of faucet added to survey question.

enjoyable to answer is to break away from the conventional layout structure by using more visually creative layouts, such as those shown in Figure 11.12.

Several published papers have reported how visually creative layouts improve both data quality and respondents' experience. Puleston and Sleep (2008) reported the following results:

- Less straightlining: up to 80% lower levels in some experiments

- Lower neutral scoring: average 25% lower

- Lower dropout rate (if questions are designed ergonomically): able to reduce from 5% to 1% in test experiments[1]

Essentially, if a survey is a pleasant experience, respondents will stay with it and think more about their answers, which reduces straightlining. Recently, GMI tested this approach on

---

[1] Egonomics is an important consideration in the design of more visual and creative question formats. Often, attempts to design more elaborate, creative question design mechanics have failed in that that some respondents do not understand how to answer them or figure out how to use them. Drag-and-drop questions are a particular problem; respondents may get stuck if instructions do not explain clearly how to move options. The foremost considerations when designing any question are ease of use and ensuring the answer process is intuitive. This often means proper testing of questions with real respondents to ensure the questions work as developers think they are going to work.

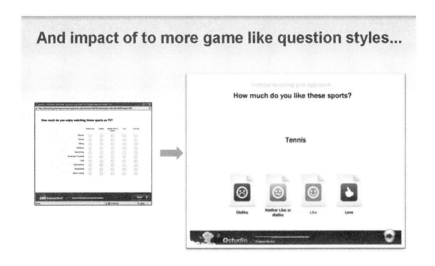

(a)

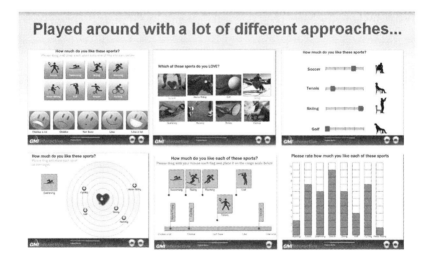

(b)

FIGURE 11.12 (a) Examples of visually creative layouts. (b) Examples of visually creative layouts.

an international scale. In the examples shown in Figure 11.13, visual icons were added to a slider to make the selection process more fun. As the respondent moved the slider, the icons changed to reinforce the selection choice.

In every country in which this approach was tested, respondents found it more fun and spent more time answering. An average 20% reduction was noted in speeding and speeding data variance and an overall increase in the standard deviation of

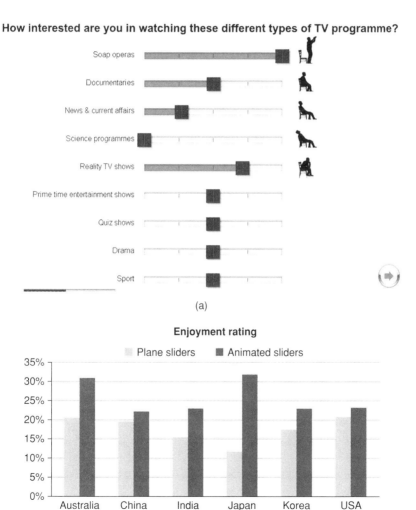

How interested are you in watching these different types of TV programme?

(a)

Enjoyment rating

(b)

FIGURE 11.13   (a) Results of addition of icons to sliders. (b) Enjoyment rating.

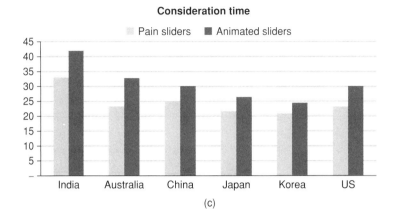

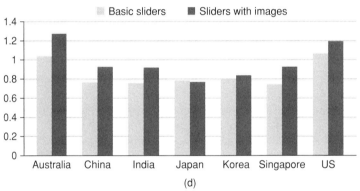

FIGURE 11.13 (*Continued*) (c) Consideration time. (d) Standard deviation.

responses (as measured at a respondent level). Much more cross-country consistency was also found in responses among countries (see Figure 11.14) (Rintoul & Puleston, 2012b).

In another example, survey questions were illustrated using "star stamp" effects. When respondents chose a selection, the choices were stamped onto the imagery (see Figure 11.15). The addition of these stamps encouraged respondents to click on nearly 50% more selections, as shown in Figure 11.16.

3. Get rid of the grid lines. A final piece of generic advice is to get rid of the grids in a survey where possible. Grids are a principal cause of dropout in surveys. At GMI, we examined the dropout points from 550 surveys and 15,000 questions, and grid questions are the cause of twice as many dropouts as any other

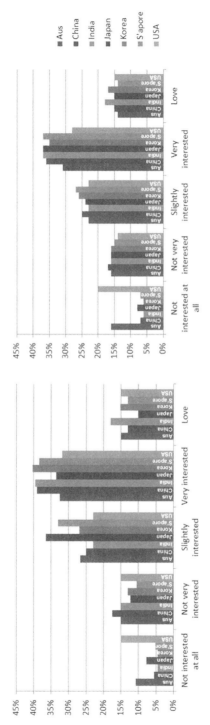

FIGURE 11.14 Cross-country consistency in responses.

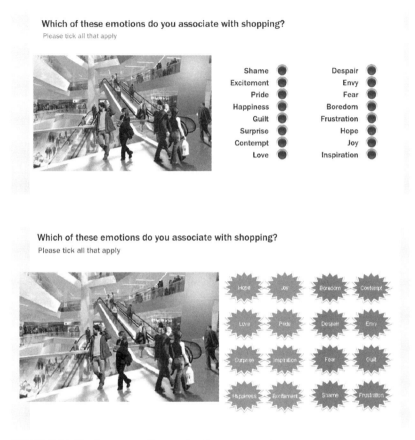

FIGURE 11.15 Star stamp effect used with survey questions.

question format. Respondents hate them. As part of the research conducted with The University of Wollongong, we looked at the time respondents spend answering different types of question in surveys. A typical respondent spends an average of 4 seconds thinking about an answer for a one-off question in a survey, but thinking time drops to under 2 seconds for questions presented in a grid (Rintoul & Puleston, 2012a). If more than 20 rows of choices are presented in a grid, by the 20th repetition, more than 80% of respondents will stop thinking and start making up answers (Rintoul & Puleston, 2012a).

Grid questions in surveys, although often used as a quick and easy way of grouping many unrelated topics into a single

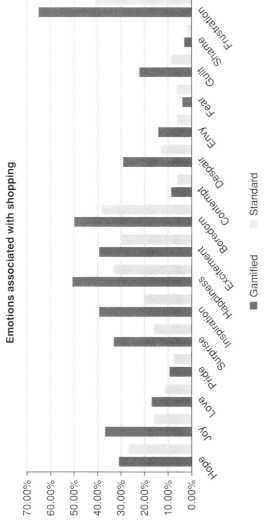

FIGURE 11.16 Gamified versus standard responses to questions.

item, should be avoided. They are especially problematic when a grid combines disparate subject matter; e.g., how much do you agree with these statements? "I like cheese," "I will vote for the Democrats," "I believe in God...," "I am going to watch the Olympics." One reason this is a problem is that a respondent's strength of opinion on one item will anchor opinions on the other topics and, therefore, may bias the survey results. If there is need to evaluate many different topics in a relative manner, I suggest presenting choices in small sets rather than lumping many topics into a single item. For example, instead of asking respondents to rate the importance of 12 different factors in a single item, present the 12 items in three random sets of four and ask respondents to pick the most and least important in each set.

COMMON QUESTIONS ABOUT GAMIFICATION

The gamification of research gives rise to three common questions: Who responds to gamification techniques? What impact does gamification have on the data? How do these techniques work in different cultures?

Who Responds to Gamification?

Many people's first thoughts are that this technique specifically applies to research on younger, more niche audiences. Although gamification is well suited to a young audience, our research has found that more than 90% and as high as 95% of respondents respond more positively to gamified question approaches included in a survey across many different demographic groups. Researchers have tested this theory by making question tasks completely voluntary and have found that applying a gamification technique consistently improves voluntary completion levels across nearly all demographic groups when applied thoughtfully.

Gamification has to be implemented in an appropriate manner, however. GMI has used avatar characters, for example, in surveys and has found that some respondents can find them trite and annoying if they look too cartoonish.

Other gaming techniques have proved a bit too difficult for respondents, most notably a skiing game where respondents had to ski downhill virtually and pass through different gates to indicate their opinions. The exercise was too difficult for some respondents and produced totally chaotic data. One can regard gamification like advertising. Ads are generally produced to improve sales, but not all ads are effective and, in some cases, may even have a negative effect. The application of gamification in surveys is a similar creative process that does not come with guarantees of success.

What Impact Does Gamification Have on the Data?

Many gamification techniques result in changes in the way questions are worded, which can significantly impact how questions are interpreted. As a result, the use of certain gamification motivation techniques is somewhat controversial, including what impact these techniques might have on the character of data.

Figure 11.17 provides examples of word clouds again from the Game Experiments paper, which exemplifies this issue (Puleston & Sleep, 2011). One group was asked to name places they liked to visit on vacation, while members of the second group were asked to imagine they were editor(s) of a travel magazine and asked to create a list of places they like to go on vacation.

Although the volume of feedback from the virtual magazine editors doubled, the answers were significantly altered by the technique used to ask the question because the emphasis of the question changed, turning it into a slightly different question. As a result, the character of the answers tended to be more exotic, which might be seen as positive or negative depending on the goals of the survey. This example illustrates the impact that wording alone can have and is a good example of the challenges survey researchers face when trying to introduce more creative questions into a survey. It is often all too easy to ignore the subtle ways that the task is being changed. Adopting these techniques brings about

FIGURE 11.17   Vacation word clouds: personal versus editors' perspectives.

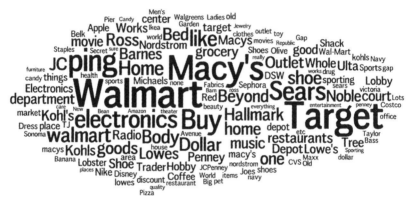

FIGURE 11.18 Word clouds of favorite stores.

two challenges: (1) determining what will motivate respondents and (2) figuring out how to implement creativity without affecting the character of the answers.

Another example comes from a follow-up experiment where one group of people was asked, "Please make a list of your favorite stores," whereas for the second group, the question was phrased, "Imagine you could design your perfect shopping center only with the stores you wanted. Can you draw up a list of the stores you would have in your shopping center?" See Figure 11.18 for the contrast in results.

Again the game mechanic changes the question somewhat. But in this case the revised wording technique seemed to have little noticeable impact on the character of the data but increased the volume of feedback fourfold. So in this case the gamification mechanic worked quite well.

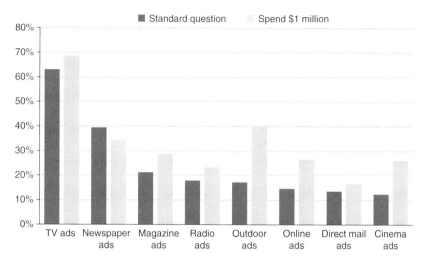

FIGURE 11.19   Media planning task from two perspectives.

Changing the question is not always a problem. The key question one should always be asking is "what am I trying to find out?"; then the answer can be used as the platform for judgment.

In a third example, 400 people were asked to undertake a media planning task. Half were asked in a traditional question what media they would choose. The second half were asked to "Imagine you had a million dollars of your own money to spend to reach people like them. What media they would choose?" As shown in Figure 11.19, respondents' choice of media changed measurably, with statistically significant changes in appeal of some less well-known media, such as billboards, online, and movie theater advertising. Perhaps these results occurred because they thought they had more money to spend, but for whatever reason, the question changed the answers.

The results beg the question, which data are more reliable? It may or may not help to note that the second group spent 50% more time thinking about their answers.

A fourth example was an experiment where respondents were asked to play a game in which they had to predict the words people associated with different brands; after each question, interviewers told them whether they were right or wrong. What slowly happened was that respondents learned what the most popular words were and began to, in effect, cheat. This effect badly corrupted the data. See Figure 11.20

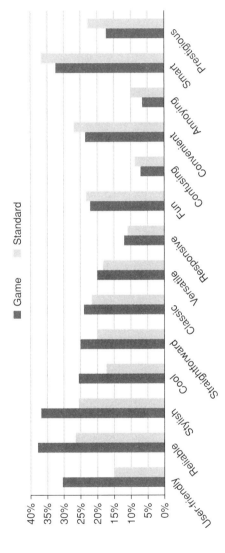

FIGURE 11.20 Standard versus game word association responses.

comparing responses with a normal word association question. The solution to this dilemma was to give respondents their score after a few questions, rather than give them the answers after every question.

These examples illustrate that gamification techniques have the potential to dramatically change the character of data—sometimes for the good, sometimes to the detriment of the data. How questions are worded and constructed depend on researchers' creative skill, but it is also a choice and clearly a critical part of survey design. Developing rules or instructions about how to use gamification in surveys is difficult. The essential tenet is to be conscious of the impact gamification techniques can have. Before using any form of gamification methodology, especially for sensitive issues, ensure that procedures for pilot testing different wording or new question styles are part of the survey development process.

How Do These Techniques Work in Different Cultures?

One common question is whether these gamification techniques work in different cultures. GMI explored this subject in a 2012 ESOMAR paper (Rintoul & Puleston, 2012b). And in another subsequent analysis (Rintoul & Puleston, 2012a), 22 different creative, gamified question techniques were tested in 14 different countries. The answer is yes— they do work in other cultures. The instinct to play seems to be a fundamental human characteristic that transcends cultural boundaries. The creative and gaming techniques that work in Western markets seem to work equally well, if not better, in the Asian markets that were tested.

GMI conducted two large-scale research experiments interviewing some 3,883 respondents in the first wave and 7,450 in the second wave in 14 different countries: Australia (613), Brazil (533), China (wave 1: 692; wave 2: 658), Germany (854), India (wave 1: 620; wave 2: 498), Japan (695), Korea (217), Mexico (484), Russia (531), Singapore (552), South Africa (499), Spain (491), Sweden (297), United Kingdom (830), and United States (wave 1: 514; wave 2: 1,776).

In total, 22 different creative gamification research techniques were tested. In nearly every case, improvements in consideration time and respondent engagement were found in all 15 countries, which is similar to those achieved in the original game experiments conducted in the United Kingdom and the United States.

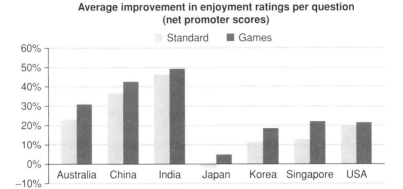

FIGURE 11.21 Average improvement in enjoyment ratings per question (net promoter scores) (standard versus games).

Figures 11.21–11.23 show examples from the first wave of this international research. Overall enjoyment scores increased in parallel to answer consideration times, improved data granularity, and lowered levels of straightlining (as measured by the average standard deviation of responses to each question by each respondent) and (where measured) volume of feedback generated. They also helped, to some degree, reduce cross-country response biases, a common problem when conducting international research. Respondents in different countries answer questions in often dramatically different ways for a variety of cultural reasons. When respondents get bored answering surveys, they also respond in different ways; for example, disengaged Japanese

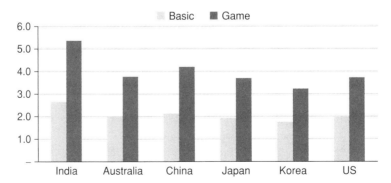

FIGURE 11.22 Average improvement in consideration time per question (seconds) (basic versus games).

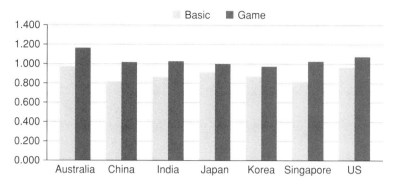

FIGURE 11.23 Average standard deviation in responses per respondent (basic versus game).

respondents have a habit of being neutral about things, and at the other end of the scale, the disengaged Indians tend to pick answers in the top end of the scale. In our research, we found that adopting more creative and gamified question approaches improved respondents concentration and reduced this noise caused by respondent disengagement. This by no means eradicated cross-country differences, but in our experiments we were able to demonstrate roughly reductions in intracountry variance by one third.

CONCLUSIONS

Gamification is transforming how and what people think about constructing market research studies. It is a highly versatile concept that can be applied on many levels in the survey design process, be it in tweaks to question wording or full-scale gamified survey experiences.

For the last 10 years, GMI has been exploring ways of making better online surveys. Gamification is the most powerful technique we have used as a means of stimulating more feedback and better quality feedback. However, it is just one of many effective techniques that can improve surveys. The point is to use gamification judiciously, not necessarily for every survey. An analogy is using humor to engage with people; it works very well but is not appropriate in all circumstances. The main takeaway is that a strong relationship exists between the level of fun respondents experience in a survey and the effort respondents put into completing a survey (see Figure 11.24).

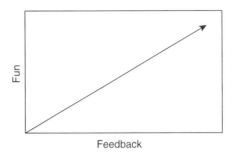

FIGURE 11.24 Level of fun versus respondents' level of effort.

More information about this topic is available at http://gamification .org/, the most comprehensive wiki site with examples of these techniques from across the fields of marketing, entertainment, education, and beyond. Theory can be helpful, but at the end of the day, one cannot escape the fact that survey design is a creative process that requires a different level of thinking.

REFERENCES

British Broadcasting Company (BBC). (2010, January 22). *Gorillas 'ape humans' over games*. Retrieved from http://news.bbc.co.uk/1/hi/scotland/edinburgh_and_east/8474358.stm.

Gartner (2011). *Gamification wiki*. Retrieved from http://gamification.org/.

KOECHLIN, E., CORRADO, G., PIETRINI, P., & GRAFMAN, J. (2000). Dissociating the role of the medial and lateral anterior prefrontal cortex in human planning. *Proceedings of the National Academy of Sciences of the United States of America*, *97*(13), 7651–7656. Retrieved from http://www.ncbi.nlm.nih.gov/pmc/articles/PMC16600/.

MAESTRIPIERI, D. (2012). *Games primates play: An undercover investigation of the evolution and economics of human relationships*. New York: Basic Books.

PULESTON, J., & SLEEP, D. (2008). *Measuring the value of respondent engagement*. Presented at the 2008 European Society for Opinion and Market Research (ESOMAR) Conference.

PULESTON, J., & SLEEP, D. (2011). *The game experiments: Researching how gaming techniques can be used to improve the quality of feedback from online research*. Presented at the 2011 ESOMAR Conference.

RINTOUL, R., & PULESTON, J. (2012a). *Can gaming techniques cross continents?* Presented at the 2012 ESOMAR Conference.

RINTOUL, R., & PULESTON, J. (2012b). *Dimensions of data quality*. Presented at the 2012 ESOMAR Conference.

Sleep, D., & Puleston, J. (2009). *Panelist engagement: Leveraging interactivity to fight boredom in online surveys*. Presented at the Advertising Research Foundation 55TH Annual Convention & Expo, New York, NY.

The Independent (2012, November 12). Human intelligence peaked thousands of years ago and we've been on an intellectual and emotional decline ever since. Retrieved from http://www.independent.co.uk/news/science/human-intelligence-peaked-thousands-of-years-ago-and-weve-been-on-an-intellectual-and-emotional-decline-ever-since-8307101.html.

Wikipedia (2012). *Games*. Retrieved from: http://en.wikipedia.org/wiki/Game.

CHAPTER 12

The Future of Social Media, Sociality, and Survey Research

Craig A. Hill and Jill Dever, *RTI International*

In Chapter 1, we discussed the troubled waters in which the survey research ship currently sails, and how the approaching hurricane of social media threatens to capsize that venerable vessel. We introduced our sociality hierarchy as a possible port in the storm, suggesting that survey researcher-captains consider making necessary repairs and improvements to the boat and pointing their sails to catch the gale force winds of social media. But, the saltier sea captains among our ranks will decry social media as only a siren song, tempting us to steer our vessels to be dashed against the rocks of the latest fad. This social media whirlpool is a vast, undecipherable mist, one might say, and is not representative of the entire ocean. It is an eddy, ceaselessly circling, with no beginning and no end. It will sink our ship or, at the very least, render it not fit for use.

In this final chapter, we offer some sturdy rope as a lifeline. Here, we lay out our vision of the future, delineated by the three levels of our social hierarchy: broadcast, conversational, and community. We will show how survey researchers can use and embrace these data and methods. But

Social Media, Sociality, and Survey Research, First Edition.
Edited by Craig A. Hill, Elizabeth Dean, and Joe Murphy.

before we do, we would be remiss if we did not address, at least in part, the "representativeness" question: As we consider incorporating social media data into our work going forward, must we abandon the very principles on which modern survey research was founded? What are the statistical challenges inherent in using social media data?

STATISTICAL CHALLENGES WITH SOCIAL MEDIA DATA

More is better! Ask a 6-year-old staring at the cookie jar (or a fisherman staring at the sea), and each might agree that more is better. The same holds for many researchers with respect to data from social media sources—*more data mean that "the universe" that one wants to study can be completely explained through Big Data and the computer processing speed for analysis makes this possible*. However, are more data truly better?

In this book, we have examined the current utility of social media sites such as Facebook, Twitter, Second Life, and the like. We have described studies carried out to conduct trend analyses (trends of Tweet sentiment obtained through Twitter discussed in Chapter 3), experiments (randomized response technique study in Chapter 6), evaluating instruments through cognitive testing (of instruments using Skype discussed in Chapter 5), and feasibility testing (recruitment of Second Life citizens with chronic health conditions discussed in Chapter 10). In other words, only estimates from the sample were desired. Survey researchers, however, use these pilot studies as guides for the implementation of a larger survey with the goal of producing population estimates. Hence, the future for social media data will include an ever expanding role in generating population estimates.

Could responses obtained from social media, collected quickly and cheaply, replace a traditional survey? Survey sampling and estimation rely on the well-documented theory developed by such experts as Hansen et al. (1953a, 1953b); Cochran (1977); Kish (1965); Särndal et al. (1992); Valliant et al. (2000); and Rao (2003), to name a few. For many survey statisticians and methodologists, these theorists not only provide instructions for drawing statistical samples, but also they represent what many consider a "good" survey. The following brief section on statistical challenges is a refresher for researchers about the basics of survey sampling and estimation in which we discuss the hurdles that must be overcome with this new source of data. And we acknowledge that yet to be developed is the theory that would allow us to expand beyond the boundaries of the sample into the unknown.

Quality and Representativeness

In survey and social science research, studies are often judged by the statistical quality of the results. Low-quality studies generally produce little more than an initial glance from the research community. They even may garner the label "a total waste of time and money." Quality, however, is defined in many ways and contains a variety of metrics such as response rates, nonresponse bias, measurement error, and the like. As we note in Chapter 1, a commonly accepted definition of quality is data that are "fit for use," in short, data that are accurate, timely, and accessible (Biemer & Lyberg, 2003). To date, the accuracy component of this three-part definition typically means whether we can justify the claim that study estimates are representative of the population of interest.

Within the survey research context, the sample is said to be representative (i.e., accurate) if it mirrors the target population and produces estimates for this population with little or no bias. As defined by Lohr (2010) and used here, a target population is "the complete collection of observations" under study defined prior to obtaining the survey sample. It differs from the sampling frame that, at least in theory, contains an enumeration of every unit in the target population. The population could be static, as in all U.S. adults ages 18 to 24 during a specified 6-month period, or conceptual, as in a subset of the example static population actively engaged in a debate on a recent world event.

Representativeness can be demonstrated through responses to a series of questions. Some examples include the following four questions, two for the sampling phase of the study and two for the estimation phase. Key words are given in parentheses to link the questions to the discussions below.

Question 1 (*Coverage Error*): How well does the sampling frame (i.e., source of the sample) cover the population under study, both overall and within important subgroups critical for the analyses?

Question 2 (*Random Sampling*): Was the sample drawn using a random methodology that is reproducible?

Question 3 (*Unbiasedness*): Do the estimates contain only negligible levels of bias suggesting that the responding sample aligns with the population?

Question 4 (*Precision*): Is the precision for the set of important estimates above some predefined levels, thereby indicating an adequate sample size for the analyses?

On the other hand, how is quality defined in the social media context? Under the fitness-for-use definition, these data are timely and accessible, sometimes within seconds of someone hitting the ENTER key. The third component, accuracy, as mentioned in Chapter 1, is more difficult to defend.

Below, we discuss accuracy from the perspective of traditional survey sampling and use this to inform concerns for Big Data. We then discuss issues related to estimating population values. Where appropriate, we borrow from current literature on the uses of nonrandom samples for survey research.

Sampling from Social Media Sources

Särndal et al. (1992) give a rigorous definition of probability sampling (and hence design-based theory) from a finite population of size N, where N is a sometimes large but quantifiable number. Of particular import in this definition is the use of a specified random sampling procedure with reproducible results, and that all members of the target population obtained from one or more sampling frames have a known nonzero probability of selection. Using the selection probabilities, design weights for sample units—also known as base weights or sample weights—can be computed. Design weights are one vehicle that researchers use to project the random survey sample to the target population. Other approaches are discussed in the following section.

Sampling frames for social media can come from a variety of sources. For example, Twitter accounts are established through only one vendor, and this administrative list of users could form the sampling frame for any point in time. The same holds true for Facebook, LinkedIn, Skype users, and any other single-source social media site. Depending on the level of detail contained in the frame, samples may be drawn just like any survey with methods such as stratified simple random sampling or probability proportional to size sampling after determining the minimum sample size using optimal allocation techniques and power calculations.

The quality of the sampling frame for correctly capturing the target population (Question 1 under the Quality and Representativeness section of this chapter) can be described by at least three characteristics (Groves, 1989; AAPOR, 2011):

1. *Undercoverage*—a portion of the target population is excluded from the estimation process. Undercoverage may occur if the sampling frame does not contain all members of the population (e.g., out-of-date information that does not capture new members) or the sampling methodology precludes certain domains (e.g., not all teens ages 16–18 in the United States use Facebook). See Figure 12.1 for a visual description.

2. *Overcoverage*—units that are not part of the target population (i.e., ineligible) are unknowingly sampled for the study. Known ineligibles can be removed from the sample after data collection and can lower the precision of the estimates. Otherwise, their influence on the bias of the estimates depends on their relative size in the sample. This could occur, for example, if e-mail addresses are sampled to access those with Twitter accounts and some accounts close just prior to the survey. See Figure 12.1 for a visual description.

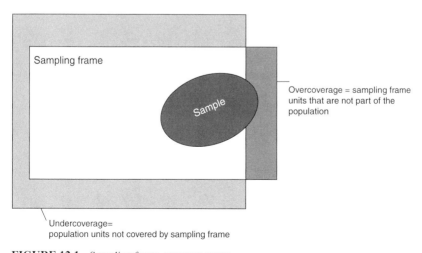

FIGURE 12.1 Sampling frame coverage errors.

3. *Multiplicities*—the same target population members are accessed more than once. For example, persons may have more than one Skype account, Twitter account (see the experiment discussed in Chapter 9), or avatar in Second Life. The same discussion given for overcoverage also applies here provided that information can be obtained to identify the duplicates.

Researchers designing the study should understand the linkage between the target population and the sampling frame from which the sample was selected. This enables the study team to explain what portion of the population is not covered by the study, or to demonstrate how the estimates are not affected by the excluded portion. Additionally, frame errors may be estimated by comparing the known information on the sampling frame with other sources linked to the target population, thereby giving a glimpse of the quality of the study estimates. Conversely, the link between the target population and the respondent estimates from a nonprobability sample is less well defined or even nonexistent. Given this, the defense of estimation techniques for the nonprobability designs requires a stronger argument and defensible assumptions.

Why does it matter if the sampling frame does not completely cover the target population? The answer is because of *bias*. Bias[1] is defined as the difference between the sample estimate and the population value such as

$$Bias(\theta) = \hat{\theta} - \theta$$

where $\hat{\theta}$ (using the hat notation) is a sample estimate such as $\bar{y}$, the mean of some characteristic y; and θ (sans hat) is the corresponding population value such as $\bar{Y}$, the population mean. As the true population mean is unknown (if it *were* known, one would not conduct the study), bias is sometimes estimated with a population mean obtained from an outside source such as from a larger survey or from historical data. The goal is to produce estimates with small levels of bias, where small can be defined in either substantive terms (i.e., the difference is not meaningful) or statistical terms (i.e., hypothesis tests indicate that the difference is not

---

[1] For studies using a sample that was not drawn using a random probabilistic method, bias is often referred to as estimation error as a distinction (OMB, 2006).

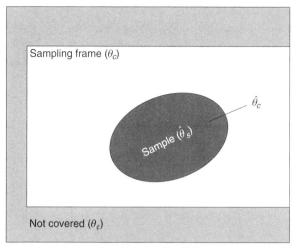

Target population (θ_U)

FIGURE 12.2 Sampling frame coverage errors and sample estimates.

statistically different from zero given a specified level of significance) (see, for example, a discussion on nonresponse bias analysis in Olson, 2006).

So, returning to the concept of undercoverage (see Figure 12.2), if the population estimate for the covered portion of the target population (θ_c) is the same as the target population value (θ_U), coverage error does not exist and the sampling frame is fine. If the sample estimate $\hat{\theta}_s$ can be used as (or augmented to produce) an accurate estimate $\hat{\theta}_c$ for the covered portion of the target population, then the sample estimate is unbiased. For example, if researchers obtain responses through Facebook, then this sampling frame and hence the estimates would exclude approximately 34% of online U.S. adults (Pew Internet & American Life (2012). If the information of interest is the same for those visiting and not visiting Facebook, then there are no worries regarding undercoverage.

Conversely, as bias increases either because the sampling frame is faulty and/or the sample estimate is not accurate, so does the likelihood of producing erroneous results that could lead to false interpretations about the underlying population. As the percentage of population units not covered by the sampling frame increases, so does the worry that bias will affect the estimates and the reliance on adjustment methods to correct for it. The next section reviews some of those methods.

What about those situations where a sampling frame is not easily developed (violating "timeliness") or when the frame does not exist (violating "accessible")? For example, the sampling frame or frames for bloggers who might comment on a particular topic, persons who visit a certain type of website, or even a group with a vested interest in (and opinions on) a recent world event are less well defined or even nonexistent. Without a sampling frame and a link to the target population, one must rely on nonprobabilistic methods such as those developed for market research and opt-in panels. Their sampling toolkit differs from those of the random probability samplers and includes the following techniques. All of these strategies are available for sampling from social media sites.

- *Quota sampling* (Moser, 1952; Moser et al., 1985)—study participants are chosen in proportion to characteristics in the population or in relation to rates established for an analysis plan. This is a specific type of *sample matching* technique and requires detailed information on the potential study participants. Examples include mall intercept surveys where sample members are chosen based on easily identifiable demographic characteristics and location within the facility.

- *River sampling* (Baker et al., 2010)—study participants are recruited only through certain websites chosen to maximize coverage of the target population. The participants are typically recruited into a panel from which samples are drawn for a specific study. The panel, therefore, acts as a sampling frame even though the frame is built using nonprobability methods. Examples of opt-in panels can be found at the GfK Knowledge Networks website.[2]

- *Network sampling* (Goodman, 1961; Granovetter, 1976; Thompson & Seber, 1996)—secondary study participants are identified through social network information collected from an initial set of study participants. The initial sample members are selected either through probability or nonprobability methods. Note that network sampling goes by many names including *respondent-driven, snowball*, or *adaptive sampling*.

---

[2] http://www.knowledgenetworks.com/.

Unlike techniques that include random sampling, nonprobability sampling is not currently based on a working theory. Instead, this approach requires more assumptions than its probability counterparts to strengthen the claim of representativeness and, hence, its fitness for use. Estimation techniques used in an attempt to address the theoretical gaps are discussed in the next section.

Population Estimation from Social Media Data

Survey estimation falls into one of three groups: design-based, model-based, and model-assisted. Each is briefly discussed below.

Design-Based Estimation

Probability sampling theory is the basis for design-based estimation (also known as randomization theory) and requires an explicit form for the probability of inclusion into the sample. Design weights (d_k), calculated as the inverse of these probabilities, are used to calculate a population estimate such as a population mean, $\hat{\bar{y}} = \left(\sum_s d_k y_k \right) / \sum_s d_k$, where y_k is the value y for the kth sample unit in the sample s; $\hat{t}_y = \sum_s d_k y_k$, which is the estimated total of the y variable in the population; and $\hat{N} = \sum_s d_k$, which is the estimated size of the population. The design weights are further adjusted to reduce the biasing influence of the errors discussed above. We label the final adjusted design weight as the analysis weight.

Adjustments are made to the design weights to reduce biases known to affect the estimates. Three biases associated with the sampling frame were noted previously. Two postsampling biases of note are:

- *Noncontact*—not all sample members can be contacted to participate in the study. Noncontact can occur when either no (or inaccurate) contact information is obtained for a sample member or when the sample member does not respond to the contact method (e.g., a person visits a feeder webpage but does not enter the study recruitment site).

- *Nonresponse*—not all contacted sample members agree to participate in the study. For surveys without a defined sampling frame, the magnitude of nonresponse and information on the nonrespondents are unknown.

Several weight adjustment techniques are in use in an attempt to adjust for sampling and postdata collection biases (see, e.g., Kalton & Flores-Cervantes, 2003).

Model-Based Estimation

The model-based perspective differs from design-based theory because it relies on the assumption that the model to describe the random variable y is correctly specified along with the underlying distribution of the variable itself instead of the randomization mechanics and associated design weights. Like mathematical statistics, the samples are assumed to be a random representation of the target population (Casella & Berger, 2002). Under simple random sampling, no weights are needed to produce ratio estimates such as means. For more complicated estimators or more complicated designs, the random assumption is strengthened by the use of mathematical models that contain the key covariates including those associated with the probabilities of inclusion.

Model-Assisted Estimation

An important and sometimes criticized assumption for model-based analyses is that the model is correct or close enough to the true population model. Särndal et al. (1992) coined the phrase model-assisted estimation to describe estimates that incorporate weights and borrow strength from (but are not dependent on) a model that incorporates strong covariates. Because weights are used, many still classify this approach as a design-based methodology. And just like design-based estimation, the design weights are adjusted to account for (potential) biases that affect the sample estimates. Two well-known and widely used examples of weight adjustments (weight calibration in particular) are the *poststratification* and *raking-ratio methods*, which are specialized forms of a generalized regression estimator (see, e.g., Kalton & Flores-Cervantes, 2003; Särndal et al., 1992). If the model variables include those associated with coverage and nonresponse, then the calibration adjustment has been shown to reduce such biases that lower quality (Kott, 2006).

Provided that random (probability) samples are selected from the social media environment, any of the approaches discussed above would be valid provided that the assumptions discussed above are met.

However, as discussed previously, not all situations fit neatly into the random sampling category. Again, we examine the literature for nonprobability sampling as a guide.

Survey researchers working on nonprobability studies have developed an estimation toolkit based on the probability sampling procedures discussed above. At the forefront are techniques to lower bias. Five sources of bias have been presented for probability surveys: undercoverage, overcoverage, multiplicities, noncontact, and nonresponse. The distinction among these sources of error is not so apparent with nonprobability surveys. Instead, *selection bias*, the source for non-negligible estimation error, is defined as the combination of all sources of error discussed previously in addition to any additional errors associated strictly with nonprobability sampling (Lee & Valliant, 2009).

Estimation procedures for nonprobability surveys attempt to reduce selection bias (and improve accuracy) and can be classified into three general categories that mirror those discussed for probability surveys:

- *Pseudo–design-based estimation*—as with design-based estimation for probability sampling, weights can be used to produce some estimates for nonprobability surveys. The term "pseudo" is used here because, unlike traditional design-based estimation, the selection probabilities are unknown and must be estimated. Propensity score adjustment (PSA) is one of several weighting methods that attempts to remove bias by balancing known sample characteristics or responses with those obtained from a high-quality reference (probability) survey (Rosenbaum & Rubin, 1984; Lee & Valliant, 2009; Bethlehem et al., 2011). As noted in Valliant and Dever (2011), bias is not corrected for any nonprobability survey variable associated with the likelihood of response that is not also included on the reference survey instrument. This finding may shed light on the mixed results on levels of bias found in the research to date (Malhotra & Krosnick, 2007; Loosveldt & Sonck, 2008; Yeager et al., 2011).
- *Model-based estimation*—model-based estimation in general relies on "healthy" covariates such as those used in a PSA that are not associated with selection bias, as well as the correct (or at least adequate) specifications of models. The Heckman two-step model is one such example (Heckman, 1979).

- *Pseudo–model-assisted estimation*—this estimation approach combines aspects of pseudo–design-based and pseudo–model-based estimation. For example, Elliott and Haviland (2007) describe a methodology for combining probability (reference) and nonprobability samples into a single data set used for modeling. Their estimator is similar to methods used for Bayesian estimation where estimates are combined in a way that gives more weight to the more precise estimate (see, e.g., Rao, 2003).

These new waters are uncharted in terms of statistical inference, representativeness, and generalizability. For some, the voyage may end here. For those brave souls who would like to venture on and envision a future, albeit incomplete, we share our thoughts on the opportunities ahead.

Future Opportunities

In the survey research world, we all deal with the fear (or the adventure) of the unknown. How can we make better population estimates without breaking the bank and without knowledge of the true population values? Probability sampling was once known as the panacea for problems linked to conducting a census where all, not a sample, of the population units were surveyed. However, as response rates continue to decline, the random nature of probability samples and the availability of funding needed to conduct these surveys are waning.

Research on sampling and estimation for probability and nonprobability samples has grown and will continue to expand, to the advantage of research using social media and other sources of timely and accessible data. The best way to introduce probability sampling into the social media paradigm may be to do research that further quantifies the coverage errors associated with the available sampling frames generated for social media (say, beyond demographic characteristics). With good quality sampling frames, standard random sampling and estimation techniques are readily available. Next, we would turn to the use of weight calibration using "social" questions included on sister (instead of on reference) surveys (see, e.g., Dever & Valliant, 2010).

Without viable (or timely and accessible) sampling frames, researchers must rely on nonprobability sampling. The sensitivity of the population estimates from social media will be tested as well as the general survey research community's sensitivity toward the use of nonprobability sampling. The best candidates for representative non-probability estimates would be the development of social media models that tap not only the currently available data but also the covariate structures that have proven the test of time.

Regardless of the approach, researchers in this field must be open and honest with their assumptions.

- Were the data collected only for pilot tests and, therefore, not generalizable?
- Are the data used to generate estimates assumed to be randomly sampled from the population of interest? If so, the researcher should clearly define this population and the mechanism with which the respondents were obtained.
- Were the assumptions underlying the estimation models evaluated? If so, the researcher should plainly describe how the assumptions were validated.

As more of these data collection efforts from social media sites are conducted, the unknown has the potential to become clearer. Therefore, in addition to validating assumptions, the quality of the study and the associated estimates may be strengthened through comparisons with historical data.

WHAT DOES THE FUTURE HOLD?

As survey researchers, we have dedicated our professional lives to producing high-quality data—data that are fit-for-use and data that we have defined as accurate, timely, and accessible. We design our studies, or surveys, so that the data collected measure population-of-interest characteristics that, in turn, allow us to estimate population-of-interest parameters. These estimates afford us, or our clients, some confidence in reaching a reliable and valid conclusion or decision based on an interpretation/analysis of these estimates.

Can we do this using social media (i.e., organic or "Big") data, either as a supplement to or, eventually, perhaps, as a replacement for survey research (i.e., designed data)? There are certainly challenges to be overcome, but the preceding chapters have shown a glimpse of what is possible, now and in the future. We are nearing the era of ubiquity—that day when every gas station within 3 miles of your GPS-generated, accelerometer-adjusted, geolocation will "know" that your car is nearly out of fuel and send to the video screen in your car its prices and wait times. Clearly, social science/survey research is on the verge of that subwave of the Information wave—one that is data driven. We do not mean to suggest that scientists now abandon theory creation and hypothesis testing and instead just surf/mine data until answers "appear" in the patterns. Instead, we mean that we want to become open to using tools and methods from other disciplines and become conversant in data management, data mining, data visualization, and data *science*.

As we evolve into the survey researchers of the future, how will we separate the noise from the signal in the data available to us? We will need to become adept at gathering data in different ways than we do now, and we will need to arrive at a deeper understanding of these data: their sources, their limitations, and their error properties. As we have shown, these data are timely and accessible, but, in short, there is much work to be done to make these data as accurate as possible.

Despite some uncertainty about the sources, limitations, and error properties of these data, researchers are already using these platforms to collect data. In recognition of the coming (sub)wave, the next section offers our thoughts about how the future of these platforms and data might play out. Figure 12.3 summarizes several examples of social media platforms, data, and methods for each level of the sociality hierarchy.

Sociality Hierarchy Level 1: Broadcast

We defined this level of the sociality hierarchy as being at the individual level, a one-to-many kind of communication in which an individual broadcasts a message to many others all at the same time. Today, research is conducted at this level by designing data capture methods that receive and analyze "transmissions" from individual social media users who provide content to a social media platform for other users to see and consume.

Currently, Twitter is the best example of this kind of broadcast transmission. Twitter's platform is elegantly simple, helping to account

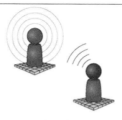

1: Broadcast

Social media platforms/data	Current Methods	Future Methods
• Twitter • Craigslist • Google	• Screen-scraping • Twitter stream sampling • Nowcasting • Infoveillance • Sentiment analysis • Google insights	• Passive data collection • Epidemiologic monitoring • Election forecasting • Detection of emerging trends

2: Conversational

Social media platforms/data	Current Methods	Future Methods
• Tumblr • Twitter • SMS (text) • Facebook • Snapchat • Instagram • Amazon mechanical turk • Online panels	• Recruitment • Registries • Tracing • Content analysis • Crowdsourcing	• Respondent-driven sampling • Instant data collector workforce • Diary/real-time data collection

3: Community

Social media platforms/data	Current Methods	Future Methods
• Second life • World of warcraft • Facebook • Nextdoor	• Cognitive interviewing • Methodological research in a controlled environment	• Building research communities (ala reconnector) • Hard-to-reach populations • "Virtual" survey laboratory • Social network analysis

FIGURE 12.3 Current and future uses of social media by sociality hierarchy level.

for its tremendous growth in a short period of time. Twitter's ground rules allow anyone to connect to anyone else without consent; I can "follow" you and thus see/read all of your 140-character Tweets without obtaining permission from you. Justin Bieber did not provide active consent to his 34,459,784 (http://twittercounter.com/pages/100; 15 February 2013) followers on Twitter. In this regard, Twitter is a unidirectional platform. When Twitter account holders Tweet, the Tweet is sent to everyone who is a follower, all at the same time. If you include those who see the Tweet as the result of a search, the total number is more than just active followers.

Probably the best-known example of using broadcast social media data to reach social-science–type conclusions is referred to as now-casting, also sometimes derisively called predicting the present. So, for instance, one can show a correlation between search queries for individual company names and the volume of stock traded for those companies (Preis et al., 2010). One of our favorite examples of now-casting is the Billion Prices Project (BPP) (http://www.pricestats.com) that began in 2008 at the Massachusetts Institute of Technology. In short, the BPP scrapes price data from about 900 different retailers in more than 70 countries, enabling its creators to calculate inflation indices in near real time. This approach not only provides data-based results that are *timely*, using *accessible* data, but it also seems quite *accurate*: The BPP results for the United States track closely with—but ahead of—the U.S. federal government's Consumer Price Index, which, of course, is expensive and requires months of data collection.

Another example of using these Level 1 data is Google Flu Trends. Somewhat famously, Google "predicted" a massive influenza outbreak in the winter of 2007–2008. By counting search queries on a particular set of keywords, such as "flu," "influenza," "flu symptoms," and the like, Google was able to track, in near real time, the rise of influenza cases in the United States. Remarkably, the rise and fall of the frequency of these search queries preceded official Centers for Disease Control and Prevention (CDC) data by about 2 weeks—using "free" data (that is, entries in the search query box on the Google homepage). Tracking of search queries, now called Google Insights, yields data that are timely, accessible, and potentially accurate (depending on the research question). Other research has shown promise for using Twitter to track flu trends as well (see Chapter 3).

Sociality Hierarchy Level 2: Conversation

Survey research has been referred to as conversation with a purpose. Survey researchers, then, can use certain social media platforms that are geared toward conversation as a way of continuing that traditional operationalization of survey research. In some ways, online panels were the first to do so. Although they do not use social media platforms, early online panel builders understood that they were, in essence, simply moving the usual way of conducting data collection to a new platform: a computer with an Internet connection. Once there, these researchers used the traditional conversation-with-a-purpose approach over the Internet (that is, asking questions and getting answers). In much the same way, survey researchers can use social media platforms to conduct a conversation with research participants. Researchers can do this quite easily or simply using, for example, short messaging service (SMS or, colloquially, texting): A researcher can send a survey question via text and then wait for an answer to be sent back. Researchers can also conduct such conversations on Facebook. A researcher can ask his "friends" a question by entering a status update or a wall post and get answers back from that set of friends. Indeed, Facebook recognizes that conversation is the bedrock of its application and has added several new features over time that leverage this, including, for example, the Invite function (that works much like asking a question: Can you come to this event? and then getting answers from a census of your friends); furthermore, now several applets allow you to conduct short "polls" of your Facebook friends.

Unlike Twitter, Facebook started out as a bidirectional platform: To "friend" someone on Facebook, both parties had to accept (actively clicking a button) the invitation. And, as people came to discover, Facebook created a problem not encountered in real life: All of one's friends are lumped together in a messy, larger-than-normal network of intermingled friends, work colleagues, grandmothers, nieces, high-school romances, and people we met at the gym once and now do not remember at all. In the real world, one does not usually have conflicting social spheres like this: If people go out for a drink with work colleagues, things said (posted) in that setting are not shared with one's grandmother and an old high-school flame at the same time.

Facebook's response was to beef up the conversational model in that one can "message" a friend directly, much like instant messaging of old or like a modern e-mail platform supported by Facebook. Facebook also now allows users to put friends in certain "bins" and to make wall posts viewable only to certain people or, conversely, to prevent a subset of friends from seeing certain posts (or even seeing one's wall at all). And, as time moved on, and as Facebook tried to become profitable, it added ways to become unidirectional: So, when one "likes," say, The Glenlivet Distillery, one sees/receives posts from The Glenlivet, but it does not see *your* posts (nor would it care, of course).

Survey researchers who wish to engage in a conversation with a purpose in the future will need to find ways to meet study participants wherever they are, which will likely mean doing it virtually. As we have seen, virtual worlds, social media sites, mobile smartphones, and other platforms in cyberspace enable researchers to conduct one-on-one conversations with a purpose.

Sociality Hierarchy Level 3: Community

At many times in this book, we have noted that survey research is being victimized by falling response rates. Survey researchers have responded in many different ways in an attempt to counteract this decline, such as using shorter surveys, providing respondents with every conceivable mode or platform to use in answering questions, or using larger and more frequent incentives (Panel on a Research Agenda for the Future of Social Science Data Collection, 2013). These attempts address only the symptoms but not the fundamental cause of the falloff in response rates.

Instead, the nature of communication, brought about by fantastic advances in communication technology, has changed enough that people—potential respondents/sample members—have a different set of expectations regarding communication with researchers (or others interested in their "data"). People, in general, enjoy the community aspects of many social media sites/platforms. They like the ability to post and share news about their lives for their friends and followers to see. They like having those friends and followers comment on what they shared and the community dialogue that occurs when more than one friend comments/shares on the original posting, commenters comment on other's comments, and so on. In effect, each person with a Facebook account, for example, has created his or her own minicommunity, composed of family members, friends, colleagues, and so forth.

Time and time again, from site to site, from social media platform to social media app, we see evidence that people are more than willing to share thoughts, insights, and pieces of their life if there is the promise of getting something of value in return—and that something of value does not have to be large or monetary; it can be as small as a virtual wink or smile from an acquaintance.

"Human societies are not simply a marketplace or an aggregation of isolated individuals seeking to exploit one another or to be free from one another. Humans are fundamentally social animals who are shaped by, and benefit from, participation in the life of the community" (Corning, 2011, page 53). We are close to seeing a networked world, a world in which we will not be asking questions of respondents/participants, but rather, one in which our "subjects" "talk back to us, and just as importantly, with each other" (Lefebvre, 2009, page 490).

Facebook and other online communities have tapped into a basic human characteristic: Disclosing information about oneself actually produces endorphins in the brain similar to eating or having sex. Additionally, even greater reward activity occurs in the brain when we know this information will be shared with family and friends (Tamir & Mitchell, 2012). The human brain is wired to enjoy forming connections, building a new neural pathway with each connection made.

As researchers, we must now think about ways to build communities that can provide us with what we value (data) while providing something of value to community participants. We have to think about ways of redefining our data collection methods so that we accommodate and leverage new norms of communication. One way to accomplish this might be to construct communities that provide support and/or interaction around a specific research topic and do this expressly, along the lines of market research online communities.

Another approach may be to use game-like interfaces that involve and engage respondents, both with the researcher and with one another. Take FarmVille (http://www.facebook.com/FarmVille) as an example: an online community built for fun that has become almost addictive to millions of players. Elements of its success include a built-in need to visit the site often, rewards for volume of activity, and the ability to trade/barter with neighbors and others in the community. Could something like FarmVille be a research platform? Could we build, say, SurveyVille, founded on these principles? Could we build SurveyVille on a platform like SurveyPulse (Chapter 7)? We would have to construct SurveyVille in such a way that the participants would want to visit often (and answer questions or, in some other way, provide data while there), be rewarded for those visits, and, probably, be able to share answers or data with other SurveyVille residents.

Or, what about Turntable (https://turntable.fm/), which has built a community of music lovers who gather together, virtually, to listen to songs and "vote" on them? Could we build, say, SurveyTable, and have a group of respondents, gathered together in the same virtual room, provide their answers/data, using the same architecture/platform as Turntable?

Another example could be Nextdoor (https://nextdoor.com/), which is building so-called private networks for neighborhoods. Could we build a SurveyDoor, centered around a "topic neighborhood" composed of members who want to chat over the virtual clothesline with others interested in the same topic?

Final Thoughts

We realize that these approaches violate some of the current principles of scientific survey research: that the researcher be objective and essentially invisible, and that respondents be anonymous. But, we are at a point in the evolution of survey research that demands new models and paradigms: "survey research is a kind of bellwether of social and technological change" (Tourangeau, 2004, page 776).

The availability of accurate, timely, and accessible data is the stuff of social scientists' dreams. "If you had asked social scientists even 20 years ago what powers they dreamed of acquiring, they might have cited the capacity to inconspicuously track the behaviors, purchases, movements, interactions, and thoughts of whole cities of people, in real time" (Christakis, 2011, paragraph 7). The advent of social media may allow the realization of some parts of this dream but not without some significant research about the best methods to collect and analyze these data moving forward.

Will social media get us to that ideal place where we can instantly read people's thoughts and do so for free and at will? Probably not—and probably that's for the best. Will social media, when viewed through the lens of the sociality hierarchy, allow us to understand our world in a way that traditional survey research has not afforded us? Maybe it will.

But if we can examine more fully, and then put into practice, what we have learned about sociality, about the fundamental changes in communication patterns and technology, and about the rapidly changing social media landscape, we have a chance to emerge with a new understanding about survey research and data science and to create the next step in the evolution of our discipline.

REFERENCES

AAPOR. (2011). *Standard definitions.* Retrieved from http://www.aapor.org/Standard_Definitions2.htm.

BAKER, R., BLUMBERG, S. J., BRICK, J. M., COUPER, M. P., COURTRIGHT, M., DENNIS, J. M., DILLMAN, D., FRANKEL, M. R., GARLAND, P., GROVES, R. M., KENNEDY, C., KROSNICK, J., & LAVRAKAS, P. J. (2010). Research synthesis: AAPOR report on online panels. *Public Opinion Quarterly*, 1–71. Retrieved from http://poq.oxfordjournals.org/content/early/2010/10/19/poq.nfq048.full.pdf+html.

BETHLEHEM, J., COBBEN, F., & SCHOUTEN, B. (2011). *Handbook of nonresponse in household surveys.* New York: Wiley.

BIEMER, P. P., & LYBERG, L. E. (2003). *Introduction to survey quality.* New York: Wiley.

CASELLA, G., & BERGER, R. L. (2002). *Statistical inference*, 2nd edition. Pacific Grove, CA: Duxbury Press.

CHRISTAKIS, N. A. (2011, June 24). The trouble with common sense. *Sunday Book Review; The New York Times.*

COCHRAN, W. G. (1977). *Sampling techniques*, 3rd edition. New York: Wiley.

CORNING, P. (2011). *The fair society: The science of human nature and the pursuit of social justice.* Chicago, IL: The University of Chicago Press.

DEVER, J. A. & VALLIANT, R. (2010). A comparison of variance estimators for poststratification to estimated control totals. *Survey Methodology, 36*(1), 45–56.

ELLIOTT, M. N., & HAVILAND, A. (2007). Use of a web-based convenience sample to supplement a probability sample. *Survey Methodology, 33*(2), 211–215.

GOODMAN, L. (1961). Snowball sampling. *Annals of Mathematical Statistics, 32*(1), 148–170.

GRANOVETTER, M. (1976). Network sampling: Some first steps. *The American Journal of Sociology, 81*(6), 1287–1303.

GROVES, R. M. (1989). *Survey errors and survey costs.* New York: Wiley.

HANSEN, M. H., HURWITZ, W. N., & MADOW, W. G. (1953a). *Sample survey methods and theory, volume I, methods and applications.* New York: Wiley.

HANSEN, M. H., HURWITZ, W. N., & MADOW, W. G. (1953b). *Sample survey methods and theory, volume II, theory.* New York: Wiley.

HECKMAN, J. J. (1979). Sample selection bias as a specification error. *Econometrica, 47*, 153–162. Retrieved from http://vanpelt.sonoma.edu/users/c/cuellar/econ411/heckman.pdf.

KALTON, G., & FLORES-CERVANTES, I. (2003). Weighting methods. *Journal of Official Statistics, 19*(2), 81–97.

KISH, L. (1965). *Survey sampling.* New York: Wiley.

KOTT, P. S. (2006). Using calibration weighting to adjust for nonresponse and coverage errors. *Survey Methodology, 32*(2), 133–142.

LEE, S., & VALLIANT, R. (2009). Estimation for volunteer panel Web surveys using propensity score adjustment and calibration adjustment. *Sociological Methods and Research, 37*, 319–343. Retrieved from http://smr.sagepub.com/content/37/3/319.full.pdf.

LEFEBVRE, R. C. (2009). Integrating cell phones and mobile technologies into public health practice: a social marketing perspective. *Health Promotion Practice, 10*, 490–494.

LOHR, S. L. (2010). *Sampling: Design and analysis, second edition.* Pacific Grove, CA: Brooks/Cole.

LOOSVELDT, G., & SONCK, N. (2008). An evaluation of the weighting procedures for an online access panel survey. *Survey Research Methods, 2*(2), 93–105.

MALHOTRA, N., & KROSNICK, J.A. (2007). The effect of survey mode and sampling on inferences about political attitudes and behavior: Comparing the 2000 and 2004 ANES to internet surveys with nonprobability samples. *Political Analysis, 15*, 286–323.

MOSER, C. A. (1952). Quota sampling. *Journal of the Royal Statistical Society,* Series A (General), *115*(3), 411–423. Retrieved from http://www.jstor.org/stable/2980740.

MOSER, C. A., KALTON, G., & MOSER, C. (1985). *Survey methods in social investigation,* 2nd edition. Surrey, UK: Ashgate.

OLSON, K. (2006). Survey participation, nonresponse bias, measurement error bias, and total bias. *Public Opinion Quarterly, 70*(5), 737–758. Retrieved from http://poq.oxfordjournals.org/content/70/5/737.full.pdf+html.

Panel on a Research Agenda for the Future of Social Science Data Collection. (2013). *Nonresponse in social science surveys: A research agenda.* Retrieved from http://www.nap.edu/openbook.php?record_id=18293&page=1.

Pew Internet & American Life. (2012). *Trend data (adults): What Internet users do online.* Retrieved from http://www.pewinternet.org/Trend-Data-(Adults)/Online-Activites-Total.aspx.

PRIES, T., REITH, D., & STANLEY, H. E. (2010). Complex dynamics of our economic life on different scales: Insights from search engine query data. *Philosophical Transactions of the Royal Society A, 368*, 5707–5719.

RAO, J. N. K. (2003). *Small area estimation.* New York: Wiley.

ROSENBAUM, P. R., & RUBIN, D. B. (1984). Reducing bias in observational studies using subclassification on the propensity score. *Journal of the American Statistical Association, 79*, 516–524.

SÄRNDAL, C. E., SWENSSON, B., & WRETMAN, J. (1992). *Model-assisted survey sampling*. New York: Springer-Verlag.

TAMIR, D. I., & MITCHELL, J. P. (2012). Disclosing information about the self is intrinsically rewarding. *Proceedings of the National Academy of Sciences*, *109*(21), 8038–8043. Retrieved from www.pnas.org/cgi/doi/10.1073/pnas.1202129109.

THOMPSON, S. K., & SEBER, G. A. F. (1996). *Adaptive sampling*. New York: Wiley-Interscience.

TOURANGEAU, R. (2004). Survey research and societal change. *Annual Review of Psychology*, *55*, 775-801.

U.S. Office of Management and Budget (OMB). (2006). *Standards and guidelines for statistical surveys*. Retrieved from http://www.whitehouse.gov/sites/default/files/omb/inforeg/statpolicy/standards_stat_surveys.pdf.

VALLIANT, R., & DEVER, J. A. (2011). Estimating propensity adjustments for volunteer web surveys. *Sociological Methods & Research*, *40*(1), 105–137. Retrieved from http://smr.sagepub.com/content/40/1/105.

VALLIANT, R., DORFMAN, A. H., & ROYALL, R. M. (2000). *Finite population sampling and inference*. New York: Wiley.

YEAGER, D. S., KROSNICK, J. A., CHANG, L., JAVITZ, H., LEVENDUSKY, M. S., SIMPSER, A., & WANG, R. (2011). Comparing the accuracy of RDD telephone surveys and Internet surveys conducted with probability and non-probability samples. *Public Opinion Quarterly*, *75*(4), 709–747.

Index

Social Media, Sociality, and Survey Research, First Edition.
Edited by Craig A. Hill, Elizabeth Dean, and Joe Murphy.
© 2014 John Wiley & Sons, Inc. Published 2014 by John Wiley & Sons, Inc.